Creating Materials with a Desired Refraction Coefficient (Second Edition)

Creating Materials with a Desired Refraction Coefficient (Second Edition)

Alexander G Ramm

IOP Publishing, Bristol, UK

ISBN   978-0-7503-3391-7 (ebook)
ISBN   978-0-7503-3389-4 (print)
ISBN   978-0-7503-3392-4 (myPrint)
ISBN   978-0-7503-3390-0 (mobi)

DOI   10.1088/978-0-7503-3391-7

Version: 20200701

IOP ebooks

British Library Cataloguing-in-Publication Data: A catalogue record for this book is available from the British Library.

Published by IOP Publishing, wholly owned by The Institute of Physics, London

IOP Publishing, Temple Circus, Temple Way, Bristol, BS1 6HG, UK

US Office: IOP Publishing, Inc., 190 North Independence Mall West, Suite 601, Philadelphia, PA 19106, USA

To Luba

Contents

Preface

The goals of this work are:

1. to give a recipe for creating materials with a desired refraction coefficient;
2. to draw the interest in solving practically (technologically) the problem of producing small particles with a prescribed boundary impedance;
3. to give a method for solving many-body scattering problem for small scatterers;
4. to prove uniqueness theorems for inverse scattering problems with *non-over-determined* scattering data;
5. to formulate a symmetry property for PDE;
6. to formulate and apply Property C for a proof of the uniqueness theorems for inverse scattering with fixed-energy potential scattering;
7. to give a method for creating materials with a wave-focusing property.

Creating materials with a prescribed boundary impedance is a problem of high interest both theoretically and practically. Such materials may be meta-materials, for example. Therefore, problem 2 is important both practically and theoretically. This is why it is discussed in detail in chapters 3 and 6. Problem 3 is basic for solving problem 1. The author's solution to this problem is given in chapter 2. In chapter 4 creating of materials with a wave-focusing property is discussed. In chapter 5 the inverse scattering theory with non-over-determined scattering data is developed for obstacle scattering. In chapter 6 creating materials with a desired refraction coefficient in bounded domains is considered for the first time. In chapter 7 a symmetry property for PDE is presented. This result allows the author to prove the Schiffer's conjecture and to solve the Pompeiu problem, see [4, 9]. In chapter 7 a new result, a symmetry property for harmonic analysis is formulated, probably, for the first time. Its proof is given. In chapter 8 the author's old result, the uniqueness theorem for inverse scattering with fixed-energy data is presented. The proof is based on the Property C introduced by the author for a study of inverse problems.

In this small book the results of the author are presented. They were published in detail in [1–9], and in the author's papers cited in the bibliography.

In this work the author presents the basic ideas and results in a self-contained way and deals with the scalar wave scattering. The emphasis is on the formulation of the results and on the importance of solving problem 2 practically.

The author thanks Momentum Press for permission to use parts of his earlier monograph [1], specifically chapters 2–4 of [1]. In chapter 5 some recent papers of the author are used [5, 6].

This book is an expanded version of the book [7], Chapters 6, 7 and 8 are added.

References

[1] Ramm A G 2013 *Scattering of Acoustic and Electromagnetic Waves by Small Bodies of Arbitrary Shapes. Applications to Creating New Engineered Materials* (New York: Momentum)

[2] Ramm A G 2005 *Inverse Problems* (New York: Springer)
[3] Ramm A G 2017 *Scattering by Obstacles and Potentials* (Singapore: World Scientific)
[4] Ramm A G 2019 *Symmetry Problems. The Navier–Stokes Problem* (San Rafael, CA: Morgan & Claypool)
[5] Ramm A G 2017 A numerical method for solving 3D inverse scattering problem with non-over-determined data *J. Pure Appl. Math.* **1** 1–3
[6] Ramm A G 2018 A uniqueness theorem for inverse scattering problem with non-over-determined data *J. Phys. A: Math. Theor.* **3** 1–5
[7] Ramm A G 2017 *Creating Materials with a Desired Refraction Coefficient* (San Rafael, CA: Morgan & Claypool)
[8] Ramm A G 2019 *Inverse Obstacle Scattering with Non-Over-Determined Scattering Data* (San Rafael, CA: Morgan & Claypool)
[9] Ramm A G 2019 Symmetry problems for the Helmholtz equation *Appl. Math. Lett.* **96** 122–5

Author biography

Alexander G Ramm

 Alexander G Ramm, PhD, was born in Russia, emigrated to the USA in 1979, and is a US citizen. He is Professor of Mathematics with broad interests in analysis, scattering theory, inverse problems, theoretical physics, engineering, signal estimation, tomography, theoretical numerical analysis, and applied mathematics. He is the author of 700 research papers, 17 monographs, and the editor of three books. He has lectured in many universities throughout the world, presented approximately 150 invited and plenary talks at various conferences, and has supervised 11 PhD students. He was a Fulbright Research Professor in Israel and in Ukraine, distinguished visiting professor in Mexico and Egypt, a Mercator Professor, an invited plenary speaker at the 7th PACOM, won the Khwarizmi International Award, and received other honors. Recently he solved inverse scattering problems with non-over-determined data and the many-body wave scattering problem when the scatterers are small particles of an arbitrary shape. Dr Ramm used this theory to provide a recipe for creating materials with a desired refraction coefficient. He gave a solution to the Pompeiu problem and proved the Schiffer's conjecture.

IOP Publishing

Creating Materials with a Desired Refraction Coefficient
(Second Edition)

Alexander G Ramm

Chapter 1

Introduction

Our method for creating materials with a desired refraction coefficient consists of distributing many small impedance particles with prescribed boundary impedance. Let us define first the notion of refraction coefficient. Let $u(x, k)$ be a wave field satisfying the equation

$$[\nabla^2 + k^2 n_0^2(x)]u = 0 \quad \text{in} \quad \mathbb{R}^3, \tag{1.1}$$

where $\mathbb{R}^3$ is the three-dimensional space, $x \in \mathbb{R}^3$, $k > 0$ is a wave number.

The coefficient $n_0(x)$ is called *refraction coefficient* of the medium.

We assume that

$$\text{Im } n_0^2(x) \leqslant 0. \tag{1.2}$$

Physically this means that the medium does not produce energy. We assume that outside of an arbitrary large but fixed area, for example, a ball $B_R = \{x: |x| \leqslant R\}$, $n_0(x) = 1$.

The basic question is:

Suppose $n_0(x)$ is the refraction coefficient of some material in $\mathbb{R}^3$. How many particles of the characteristic dimension a and the boundary impedance $\zeta(x)$ should be distributed in B_R in order to get in B_R a new material with the desired refraction coefficient $n(x)$? In section 6.2 we discuss a similar question assuming that the small impedance particles are distributed inside a closed surface Γ on which the Dirichlet boundary condition is imposed.

The wavelength λ_0 in the medium with refraction coefficient $n_0(x) = 1$ can be calculated by the formula $\lambda_0 = \frac{2\pi}{k}$. Smallness of a particle of the characteristic dimension a is described by the relation

$$a \ll \frac{\lambda_0}{n_0}, \qquad n_0 := \max_{x \in \mathbb{R}^3} n_0(x). \tag{1.3}$$

Wave scattering by the impedance particles in a medium is described by the equations

$$\left(\nabla^2 + k^2 n_0^2(x)\right)u = 0 \quad \text{in} \quad \mathbb{R}^3 \backslash D, \qquad D := \bigcup_{m=1}^{M} D_m \tag{1.4}$$

$$u_N = \zeta_m u \quad \text{on} \quad S_m, \quad 1 \leqslant m \leqslant M, \tag{1.5}$$

$$u = u_0 + v, \tag{1.6}$$

$$\frac{\partial v}{\partial r} - ikv = o\left(\frac{1}{r}\right), \quad r = |x| \to \infty. \tag{1.7}$$

Here $u_0(x)$ is the incident field. It satisfies the equation:

$$\left(\nabla^2 + k^2 n_0^2(x)\right)u_0 = 0 \quad \text{in} \quad \mathbb{R}^3, \tag{1.8}$$

the dependence on k is dropped since $k > 0$ is constant, v is the scattered field, D_m is mth small particle, S_m is its sufficiently smooth boundary (surface), N is a unit normal to S_m pointing out of D_m, ζ_m is *boundary impedance* of D_m, $\operatorname{Im} \zeta_m \leqslant 0$, M is the number of small particles. For simplicity only we assume in what follows that $n_0(x) = 1$. In chapter 8 we explain why this assumption is not important for our theory.

An important physical quantity d is defined as minimal distance between neighboring particles

$$d = \min_{\substack{j,m}} \operatorname{dist}_{m \neq j}(D_m, D_j). \tag{1.9}$$

Let us assume that

$$\zeta_m = \frac{h(x_m)}{a^\kappa}, \quad \operatorname{Im} h(x) \leqslant 0, \tag{1.10}$$

where $x_m \in D_m$, $h(x)$ is a continuous function in B_R, $\kappa \in [0, 1)$ is a number, $a = \frac{1}{2} \max_m \operatorname{diam} D_m$.

The distribution of the small particles is described by the function

$$\mathcal{N}(\Delta) = \frac{1}{a^{2-\kappa}} \int_\Delta N(x)dx(1 + o(1)), \quad a \to 0, \tag{1.11}$$

where $\Delta \subset B_R$ is an arbitrary open set, $N(x) \geqslant 0$ is a continuous function, $\mathcal{N}(\Delta)$ is the number of small particles in Δ. In particular,

$$M = \frac{1}{a^{2-\kappa}} \int_{B_R} N(x)dx(1 + o(1)) = O\left(\frac{1}{a^{2-\kappa}}\right), \quad a \to 0. \tag{1.12}$$

We prove that if the small particles are distributed by the law (1.11) then one can choose $h(x)$, that is, boundary impedances, so that the solution to the many-body scattering problem has a limit u, as $a \to 0$, and this u solves the equation

$$[\nabla^2 + k^2 n^2(x)]u = 0, \tag{1.13}$$

where the refraction coefficient $n(x)$ can be an arbitrary function satisfying the condition $\operatorname{Im} n^2(x) \leqslant 0$. Moreover, the limiting function $n^2(x)$ will be calculated explicitly

$$n^2(x) = n_0^2(x) - c\frac{h(x)N(x)}{k^2}, \tag{1.14}$$

the constant c is defined by the formula

$$|S_m| = ca^2, \tag{1.15}$$

where $|S_m|$ is the surface area of S_m, $h(x)$ is the function from equation (1.10) and $N(x)$ is the function from equation (1.11).

Our basic physical assumption is

$$a \ll d \ll \lambda, \tag{1.16}$$

where d is defined in equation (1.9) and $\lambda = \frac{2\pi}{k} \min_{x \in B_R} |n_0(x)|$. Assumption (1.16) means that multiple scattering is essential: one cannot assume that the field acting on a particle D_m is the incident field u_0.

From equation (1.14) it is already clear that if $h(x)$ can be chosen arbitrarily with $\operatorname{Im} h \leqslant 0$ then $n(x)$ can be obtained fairly arbitrarily.

In chapter 2 our theory of many-body wave scattering is presented for small impedance particles and formulas (1.13) and (1.14) are derived.

In chapter 3 a recipe for creating materials with a desired refraction coefficient is formulated and justified theoretically.

In chapter 4 we will discuss creating wave-focusing materials. These are materials such that a plane wave, scattered by such material, has a desired radiation pattern.

In chapter 5 the obstacle inverse scattering theory is developed for non-over-determined scattering data. This means that the dimension of the data is the same as the dimension of the unknown object. For the obstacle scattering the unknown object is the surface S of the obstacle which has dimension two. The scattering data we use is a function $A(\beta) := A(\beta, \alpha_0, k_0)$, where $A(\beta, \alpha, k)$ is the scattering amplitude, β is the unit vector in the direction of the scattered field, α is the unit vector in the direction of the incident field, k is the wave number, the index 0 shows that the corresponding variable is fixed. So the data $A(\beta)$ is a function of two variables β_1, β_2 and the unknown object S is described by a function of two variables.

In chapter 6 we discuss the experiments allowing us to move the refraction coefficient in a desired direction, for example, make it smaller than the original one.

We also develop a theory for creating the desired refraction coefficient assuming that the original material is located inside a bounded region on the boundary of which the Dirichlet boundary condition is imposed. This is of practical interest because our original theory, developed in chapter 2, assumes that the small particles are embedded in a bounded region and no boundary condition on the surface of this region is imposed.

The last section of chapter 6 deals with embedding the small acoustically soft particles, rather than impedance particles.

In chapter 7 a symmetry problem for the solutions to the Helmholtz equation is studied. The results allowed the author earlier to prove the Schiffer's conjecture and to solve the Pompeiu problem. In chapter 7 for the first time a symmetry problem in harmonic analysis is formulated and proved. For convenience of the reader a proof of the theorem from [1] on symmetry properties for the solutions to the Helmholtz equation is given.

In chapter 8 the inverse scattering problem with fixed-energy data is discussed. The basic uniqueness theorem, originally proved by the author in 1987 is proved.

The contents of chapters 2–5 remain basically the same as in the earlier published book [2]. In this book the new chapters 6–8 are added and chapter 1, the Introduction, is modified accordingly.

References

[1] Ramm A G 2019 *Symmetry Problems. The Navier-Stokes Problem* (San Rafael, CA: Morgan & Claypool)
[2] Ramm A G 2017 *Creating Materials with a Desired Refraction Coefficient* (IOP Concise Physics) (San Rafael, CA: Morgan & Claypool)

Chapter 2

Wave scattering by many small impedance particles

2.1 Scalar wave scattering by one small body of an arbitrary shape

2.1.1 Impedance bodies

Consider the following wave scattering problem

$$(\nabla^2 + k^2)u = 0 \quad \text{in } D' := \mathbb{R}^3 \backslash D, \tag{2.1}$$

$$u_N^- = \zeta u \quad \text{on } S, \tag{2.2}$$

$$u = u_0 + v, \tag{2.3}$$

$$\frac{\partial v}{\partial |x|} - ikv = o\left(\frac{1}{|x|}\right), \quad |x| \to \infty. \tag{2.4}$$

Here $k > 0$ is a wave number, $k = \text{const}$, D is a bounded domain with a smooth boundary S, N is the unit normal to S pointing out of D, u_N^- is the limiting value of the normal derivative of u on S from D', ζ is a constant which we call the boundary impedance, $u_0 = e^{ik\alpha \cdot x}$ is the incident field, the plane wave, $\alpha \in S^2$, is a unit vector, S^2 is the unit sphere, v is the scattered field, condition (2.4) is the radiation condition at infinity. It is assumed to be satisfied uniformly with respect to the unit vector $x^0 = \frac{x}{|x|}$, that is, with respect to the direction along which x tends to infinity. Smoothness of S throughout this book means that the equation of the surface S in local coordinates is a $C^{1,\lambda}$ function. The local coordinates are defined as follows. Fix a point $s \in S$ and let it be the origin of the coordinate system, the z-axis of which is directed along the normal N_s to S at the point s, and the plane x, y is tangent to S at this point. We often write x_1, x_2, x_3 in place of x, y, z, and then $x = (x_1, x_2, x_3)$. Let $x_3 = f(x_1, x_2)$ be the local equation of S in this coordinate system. Then, by construction, $f(0, 0) = 0$, $\frac{\partial f(0,0)}{\partial x_j} = 0$, $j = 1, 2$. The assumption $S \in C^{1,\lambda}$, $0 < \lambda \leqslant 1$, means that

$$|\nabla f(x_1, x_2) - \nabla f(x_1', x_2')| \leqslant C[(x_1 - x_1')^2 + (x_2 - x_2')^2]^{\lambda/2},$$

where C is a constant that does not depend on s, x and x', and $\nabla = e_1\frac{\partial}{\partial x_1} + e_2\frac{\partial}{\partial x_2}$, $\{e_j\}_{j=1}^3$ is the Cartesian basis of $\mathbb{R}^3$.

The scattering problem (2.1)–(2.4) can be considered, for example, as the scattering of an acoustic wave. In this case u has the physical meaning of the pressure or acoustic potential. The Dirichlet boundary condition $u|_S = 0$ describes acoustically soft body (zero pressure on the boundary) and the Neumann boundary condition $u_N|_S = 0$ describes acoustically hard body (the normal component of the velocity ∇u vanishes on S). The impedance boundary condition describes a linear relation between the pressure and the normal component of the velocity on S.

The first task is to prove that the scattering problem has a solution and this solution is unique. In this case we say that the scattering problem has a unique solution. If we want to state that there is at most one solution, but the existence of the solution is not asserted, then we say there exists at most one solution.

Theorem 2.1.1. *Assume that* Im $\zeta \leqslant 0$. *Then problem (2.1)–(2.4) has at most one solution.*

Proof. Since the problem is linear it is sufficient to prove that the corresponding homogeneous problem, that is, the problem with $u_0 = 0$, has only the trivial solution $u = 0$. To prove this, multiply equation (2.1) by $\bar{u}$, the complex conjugate of u, subtract the complex conjugate equation (2.1) multiplied by u and integrate over the region $D' \cap B_R := D'_R$, where $R > 0$ is a large number that we take to infinity and $B_R = \{x: |x| \leqslant R\}$ is a ball of radius R centered at the origin. The origin we take inside D arbitrarily. The result is

$$0 = \int_{D'_R} [\bar{u}(\nabla^2 + k^2)u - u(\nabla^2 + k^2)\bar{u}]dx$$
$$= \int_{S_R} (\bar{u}u_N - u\bar{u}_N)ds - \int_S (\bar{u}u_N - u\bar{u}_N)ds. \tag{2.5}$$

Here the Green's formula was used.

From the radiation condition (2.4) one gets:

$$\int_{S_R} (\bar{u}u_N - u\bar{u}_N)ds = 2ik \int_{S_R} |u|^2 ds + o(1), \quad R \to \infty. \tag{2.6}$$

From the impedance boundary condition (2.2) one gets:

$$-\int_S (\bar{u}u_N - u\bar{u}_N)ds = \int_S (-\zeta|u|^2 + \bar{\zeta}|u|^2)ds = -2i \, \mathrm{Im} \, \zeta \int_S |u|^2 ds. \tag{2.7}$$

From equations (2.5)–(2.7) it follows that

$$\lim_{R\to\infty}\left(k\int_{S_R}|u|^2ds - \operatorname{Im}\zeta\int_S|u|^2ds\right)=0. \tag{2.8}$$

If $\operatorname{Im}\zeta\leqslant 0$, then relation (2.8) implies

$$\lim_{R\to\infty}\int_{S_R}|u|^2ds = 0. \tag{2.9}$$

This and the differential equation (2.1) imply that $u=0$ outside any ball $B_R\supset D$. This claim is proved in [1, p 25]. Now a consequence of the unique continuation principle for a solution to homogeneous Helmholtz equation: if such a solution vanishes on an open subset in the domain D' where it solves the homogeneous Helmholtz equation, then this solution vanishes everywhere in D'.

Theorem 2.1.1 is proved. □

Remark 2.1.1. *Physically the impedance ζ can be any constant satisfying the condition $\operatorname{Im}\zeta\leqslant 0$ which guarantees the uniqueness of the solution to the scattering problem (2.1)–(2.4).*

Let us now prove the existence of the solution to the scattering problem (2.1)–(2.4).

Theorem 2.1.2. *Problem (2.1)–(2.4) has a (unique) solution. This solution can be found of the form*

$$u(x) = u_0(x) + \int_S g(x,\,t)\sigma(t)dt, \tag{2.10}$$

$$g(x,\,y) := \frac{e^{ik|x-y|}}{4\pi|x-y|}. \tag{2.11}$$

The function $\sigma(t)$ in equation (2.10) is uniquely determined.

Proof. For any σ function (2.10) solves equation (2.1) and satisfies conditions (2.3) and (2.4). Therefore, it solves problem (2.1)–(2.4) if and only if σ can be found so that condition (2.2) is satisfied, that is,

$$u_{0N} - \zeta u_0 + \frac{A\sigma - \sigma}{2} - \zeta T\sigma = 0, \tag{2.12}$$

where

$$A\sigma := 2\int_S \frac{\partial g(s,\,t)}{\partial N_S}\sigma(t)dt, \quad T\sigma := \int_S g(s,\,t)\sigma(t)dt. \tag{2.13}$$

Formula

$$\left| \frac{\partial}{\partial N_S} \int_S g(x,\,t)\sigma(t)dt \right|_{x \in D',\, x \to s} = \frac{A\sigma - \sigma}{2} \tag{2.14}$$

is proved in [2]. Equation (2.12) is of Fredholm type because operators A and T are compact in $C(S)$. Therefore, the existence of the solution to equation (2.12) follows from the Fredholm alternative if one proves that the homogeneous equation (2.12) has only the trivial solution. The homogeneous equation (2.12) is of the form

$$\frac{A\sigma - \sigma}{2} - \zeta T\sigma = 0, \tag{2.15}$$

which is

$$u_{\bar{N}} - \zeta u = 0 \quad \text{on } S. \tag{2.16}$$

By theorem 2.1.1 the corresponding function $u(x) = 0$ in D' because it solves equation (2.1) in D', satisfies the radiation condition (2.4) and the boundary condition (2.2).

If $u = 0$ in D', then $u = 0$ on S because potentials of a single layer are continuous in $\mathbb{R}^3$, see [2]. Thus, u solves equation (2.1) in D and satisfies condition (2.2). This implies that $u = 0$ in D. Indeed, multiply equation (2.1) in D by $\bar{u}$, use Green's formula and the boundary condition (2.2), and get

$$2i \operatorname{Im} \zeta \int_S |u|^2 ds = 0. \tag{2.17}$$

If $\operatorname{Im} \zeta \neq 0$, then $u = 0$ on S, so $u_N = 0$ on S and $u = 0$ in D by the uniqueness of the solution to the Cauchy problem for the Helmholtz equation. If $u = 0$ in $D \cup D'$ then $\sigma = u_N^+ - u_{\bar{N}} = 0$, and theorem 2.1.2 is proved. If $\operatorname{Im} \zeta = 0$, then one deals with the real number ζ in (2.2).

If, as we assume, the body D is sufficiently small, then k^2 is not an eigenvalue of the Laplacian in D also in the case $\operatorname{Im} \zeta = 0$. So, in this case also $u = 0$ in $D \cup D'$ and $\sigma = 0$.

Theorem 2.1.2 is proved. $\qquad\square$

The existence and uniqueness of the solution to the scattering problem does not depend on the size of the body D. This size we measure by the number $a := \frac{1}{2} \operatorname{diam} D$. We assume that the body is small in the sense $ka \ll 1$, that is, $a \ll \lambda$, where λ is the wavelength. This assumption allows us to derive analytic, closed form formulas for the field, scattered by D, at the distances $d \gg a$, in the 'far zone'.

Our next task is to derive an analytical formula for the scattered field at the distance $|x| \gg a$. To do this, take into account the following formula

$$\frac{e^{ik|x-t|}}{4\pi|x-t|} = \frac{e^{ik|x|-ikx^0 \cdot t}}{4\pi|x|} \left(1 + O\left(\frac{|t|}{|x|} \right) \right), \quad |t| \leqslant a. \tag{2.18}$$

Here $x^0 = \frac{x}{|x|}$. Formula (2.18) can be established easily. One has:

$$|x - t| = |x|\sqrt{1 - \frac{2x^0 \cdot t}{|x|} + \frac{|t|^2}{|x|^2}} = |x|\left(1 + O\left(\frac{|t|}{|x|}\right)\right) \qquad (2.19)$$

and

$$e^{ik|x-t|} = e^{ik|x|-ikx^0 \cdot t + O\left(\frac{|t|^2}{|x|}\right)} = e^{ik|x|-ikx^0 \cdot t}\left(1 + O\left(\frac{|t|^2}{|x|}\right)\right). \qquad (2.20)$$

If $|t| \leqslant a$ and a is small, that is, $ka \ll 1$, then $O(\frac{|t|^2}{|x|}) \leqslant O(\frac{|t|}{|x|})$, so equation (2.18) holds. In what follows we denote $x^0 = \beta$ and $|x| = r$. One has

$$\int_S g(x, t)\sigma(t)dt = \frac{e^{ikr}}{4\pi r}\int_S e^{-ik\beta \cdot t}\sigma(t)dt\left(1 + O\left(\frac{a}{r}\right)\right). \qquad (2.21)$$

Let $g(r) := \frac{e^{ikr}}{4\pi r}$ and

$$Q := \int_S \sigma(t)dt. \qquad (2.22)$$

If

$$|Q| \gg \left|\int_S t\sigma(t)dt\right|, \qquad (2.23)$$

then one can write

$$\int_S e^{-ik\beta \cdot t}\sigma(t)dt \sim \int_S \sigma(t)dt = Q, \qquad (2.24)$$

where *the sign $\sim$ stands for the asymptotic equality as $a \to 0$.*
    If equation (2.23) holds then

$$u(x) \sim u_0(x) + g(r)Q, \quad r = |x| \gg a. \qquad (2.25)$$

Therefore, the solution of the scattering problem is completed if the number Q is found.
    To find Q we use the exact integral equation (2.12). However, we do not solve it for σ, but find the main term of Q as $a \to 0$ asymptotically, in closed form. To do this, integrate equation (2.12) over S and evaluate the significance of each term as $a \to 0$.
    The first term is

$$I_1 := \int_S u_{0N}ds = \int_D \nabla^2 u_0 dx = -k^2 \int_D u_0 dx = O(a^3), \qquad (2.26)$$

where we took into account that

$$\left|\int_D u_0 dx\right| \leqslant \int_D dx = |D| = O(a^3), \qquad (2.27)$$

and $|D| := V$ is the volume of D.

The second term is

$$I_2 := \int_S \zeta u_0 ds = O(a^2), \tag{2.28}$$

where

$$\int_S ds = |S| = O(a^2), \tag{2.29}$$

and $|S|$ is the surface area of S.

The third term is

$$I_3 := \int_S \frac{A\sigma - \sigma}{2} ds = -\frac{Q}{2} + \frac{1}{2}\int_S A\sigma ds. \tag{2.30}$$

We claim that

$$\int_S A\sigma ds \sim \int_S A_0\sigma ds = -\int_S \sigma ds = -Q, \tag{2.31}$$

where

$$A_0\sigma := 2\int_S \frac{\partial g_0(s, t)}{\partial N_S}\sigma(t)dt, \quad g_0(x, t) := \frac{1}{4\pi|x - t|}. \tag{2.32}$$

Indeed, one has

$$A\sigma = 2\int_D \frac{\partial}{\partial N_S}\frac{e^{ik|s-t|}}{4\pi|s - t|}\sigma(t)dt$$

$$= 2\int_S \frac{\partial}{\partial N_S}\left(\frac{1 + ik|s - t| + O(k^2|s - t|^2)}{4\pi|s - t|}\right)\sigma(t)dt \tag{2.33}$$

$$= A_0\sigma + \int_S O(|s - t|)\sigma(t)dt$$

$$\sim A_0\sigma \quad \text{as } a \to 0.$$

Let us check that

$$\int_S A_0\sigma ds = -\int_S \sigma dt. \tag{2.34}$$

This is an exact relation. One has

$$2\int_S ds \int_S \frac{\partial}{\partial N_S}\frac{1}{4\pi|s - t|}\sigma(t)dt = \int_S dt\sigma(t)2\int_S \frac{\partial}{\partial N_S}\frac{1}{4\pi|s - t|}ds$$

$$= -\int_S \sigma(t)dt. \tag{2.35}$$

Here we have used the known formula (see [3], formula (1.13)):

$$2 \int_S \frac{\partial}{\partial N_S} \frac{1}{4\pi|s-t|} ds = -1, \quad t \in S. \tag{2.36}$$

Proof of this formula is given in [2], formula (11.1.7).

Thus,

$$I_3 \sim -Q, \quad \text{as } a \to 0. \tag{2.37}$$

Finally,

$$I_4 := \zeta \int_S T\sigma ds = \int_S dt\sigma(t)\zeta \int_S g(s,t) ds. \tag{2.38}$$

One has

$$\int_S g(s,t) ds = O(a), \quad a \to 0. \tag{2.39}$$

Let us assume that

$$\lim_{a \to 0} \zeta a = 0. \tag{2.40}$$

Assuming equation (2.40) we allow ζ to depend on a in such a way that equation (2.40) holds. If ζ is a constant independent of a then equation (2.40) is obviously valid. If equation (2.40) holds then

$$I_4 = o(Q), \quad a \to 0, \tag{2.41}$$

provided that $Q \neq 0$. We will prove that $Q \neq 0$ under our assumptions. Keeping the terms I_3 and I_2 and neglecting the terms I_1 and I_4 of higher order of smallness as $a \to 0$, one gets

$$Q \sim -\zeta|S|u_0(x_1), \quad x_1 \in D. \tag{2.42}$$

If $u_0(x) = e^{ika \cdot x}$ and $x_1 \in D$, then $u_0(x_1) = 1 + O(ka)$, since $ka \ll 1$.

Formula (2.42) is our *basic result*.

From equations (2.42) and (2.25) one gets an analytical formula for the solution of the wave scattering problem (2.1)–(2.4) in the case of a small body of an arbitrary shape:

$$u(x) \sim u_0(x) - g(r)\zeta|S|, \quad |x| \gg a, \tag{2.43}$$

where $g(r) = \frac{e^{ikr}}{4\pi r}$, $r = |x|$, and $u_0(x_1) \sim 1$ if $ka \ll 1$.

It is assumed in equation (2.43) that the origin is inside D. If the origin is elsewhere, and $x_1 \in D$, then formula (2.43) takes the form

$$u(x) \sim u_0(x) - g(|x - x_1|)\zeta|S|, \quad |x - x_1| \gg a. \tag{2.44}$$

Let us define the scattering amplitude $A(\beta, \alpha, k)$ by the formula

$$v \sim \frac{e^{ikr}}{r} A(\beta, \alpha, k) + o\left(\frac{1}{r}\right), \quad r = |x| \to \infty, \quad \frac{x}{r} := \beta, \tag{2.45}$$

where v is the scattered field.

Comparing equations (2.43) and (2.45) one obtains

$$A(\beta, \alpha, k) = -\frac{\zeta|S|}{4\pi} = \frac{Q}{4\pi}, \tag{2.46}$$

where it is assumed that $x_1 = 0$, so that $u_0(x_1) = 1$.

The physical conclusion is:

The scattering is isotropic, that is, it does not depend on α and β, and in absolute value the scattering amplitude is $O(\zeta|S|) = O(\zeta a^2)$.

Let us summarize the results in a theorem.

Theorem 2.1.3. *Assume that $ka \ll 1$, $\operatorname{Im} \zeta \leqslant 0$, and (2.40) holds. Then the scattering problem (2.1)–(2.4) has a unique solution. This solution can be calculated by formula (2.43) if the origin is inside D, and by formula (2.44) if the origin is elsewhere and $x_1 \in D$.*

The scattering is isotropic and the scattering scattering amplitude is $O(\zeta a^2)$.

2.1.2 Acoustically soft bodies (the Dirichlet boundary condition)

In this section problem (2.1), (2.3), (2.4) is studied but the impedance boundary condition (2.2) is replaced by the Dirichlet condition

$$u = 0 \quad \text{on } S. \tag{2.47}$$

Physically this condition in acoustics means that the body is acoustically soft, that is, the pressure on S is vanishing. Mathematically condition (2.47) is the limiting case of condition (2.2) when $\zeta \to \infty$. This is proved in [2]. The case $\zeta \to 0$ yields the Neumann condition $u_N = 0$ on S, corresponding to the acoustically hard body.

The scattering problem with the Dirichlet condition has a unique solution, as we will prove, and the analogs of theorems 2.1.1–2.1.3 will be established.

Theorem 2.1.4. *Problem (2.1), (2.47), (2.3), (2.4) has at most one solution.*

Proof. It is sufficient to prove that the corresponding homogeneous problem, that is, the problem with $u_0 = 0$, has only the trivial solution. The proof is essentially the same as the proof of theorem 2.1.1.

One obtains an analog of relation (2.8) with the integral over S vanishing because $u = 0$ on S. Therefore, one concludes that relation (2.9) holds.

The rest of the proof goes as in the proof of theorem 2.1.1.

Theorem 2.1.4 is proved. $\qquad\square$

Let us prove the existence of the solution to problem (2.1), (2.47), (2.3), and (2.4).
Theorem 2.1.5. *The above problem has a solution.*

Proof. Let us look for the solution of the form (2.10). As in the proof of theorem
2.1.2 it is sufficient to prove that the equation that one obtains from the boundary
condition (2.47) is solvable.

This equation is

$$\int_S g(s, t)\sigma(t)dt = -u_0(s), \quad s \in S. \tag{2.48}$$

It is proved in [2] that equation (2.48) is (uniquely) solvable provided that k^2 is not an
eigenvalue of the Dirichlet Laplacian in D. This condition is satisfied if D is
sufficiently small.

Theorem 2.1.5 is proved. □

Let us prove an analog of theorem 2.1.3.

As in section 2.1.1 one gets formula (2.25) where Q is defined in (2.22). To find an
analytical expression for Q we rewrite equation (2.48) as

$$\int_S g_0(s, t)\sigma(t)dt = -1 + O(ka), \quad g_0(s, t) = \frac{1}{4\pi|s - t|}. \tag{2.49}$$

Neglecting the small term $ka \ll 1$, one gets

$$\int_S g_0(s, t)\sigma(t)dt = -1, \quad s \in S. \tag{2.50}$$

This is an equation for the surface charge distribution σ that generates the constant
potential $\mathcal{U}$ on S, $\mathcal{U} = -1$. With this interpretation the body D is a perfect conductor
and the quantity $Q = \int_S \sigma(t)dt$ is its total charge. There is a well-known relation
$C\mathcal{U} = Q$, where C is the capacitance of the perfect conductor D. Thus,

$$Q = -C, \tag{2.51}$$

and formula (2.25) takes the form

$$u(x) = u_0(x) - g(r)C, \quad r = |x| \gg a, \tag{2.52}$$

while formula (2.46) becomes

$$A(\beta, \alpha, k) = -\frac{C}{4\pi}. \tag{2.53}$$

These formulas solve the scattering problem (2.1), (2.47), (2.3), (2.4).

Let us formulate the results.

Theorem 2.1.6. *If $ka \ll 1$ then the solution to the above scattering problem exists, is unique, can be calculated by formula (2.52) and the scattering amplitude is given by formula (2.53).*

Remark 2.1.2. *This result is very useful practically because the author has derived explicit analytical formulas which allow one to calculate electrical capacitance C for a conductor of an arbitrary shape, see [2] and [3].*

For example, the zeroth approximation of C is given by the following formula

$$C^{(0)} = \frac{4\pi |S|^2}{\int_S \int_S \frac{dsdt}{|s-t|}}, \qquad C^{(0)} \leqslant C, \tag{2.54}$$

where $|S|$ is the surface area of S and the electric permittivity in D' is assumed to be equal to 1.

Let us draw some physically interesting conclusions from theorem 2.1.6.

Note that the electrical capacitance $C = O(a)$. Formula (2.53) shows that the scattering is isotropic and the scattering amplitude is $O(a)$, that is, it is *much larger* than in the case of the impedance boundary condition. If, for example, $\zeta = O(\frac{1}{a^\kappa})$, $0 \leqslant \kappa < 1$, so that condition (2.40) is satisfied, then formula (2.46) yields $A(\beta, \alpha, k) = O(a^{2-\kappa})$, and $O(a^{2-\kappa}) \ll O(a)$ if $\kappa < 1$ and $a \to 0$.

2.1.3 Acoustically hard bodies (the Neumann boundary condition)

Consider now the scattering problem (2.1), (2.3), (2.4) and replace condition (2.2) by the Neumann boundary condition

$$u_N = 0 \quad \text{on } S. \tag{2.55}$$

Our plan of the study is unchanged: we want to prove existence of a unique solution and derive analytical formulas for the solution and for the scattering amplitude.

Theorem 2.1.7. *Problem (2.1), (2.55), (2.3), (2.4) has at most one solution.*

Proof. One can use the proof of theorem 2.1.1 taking $\zeta = 0$ in this proof.
Theorem 2.1.7 is proved. ◻

Theorem 2.1.8. *The scattering problem (2.1), (2.55), (2.3), (2.4) has a solution.*

Proof. Let us look for the solution of the form (2.10). As in the proof of theorem 2.1.2 it is sufficient to prove that equation (2.12) with $\zeta = 0$ has a solution. Since equation (2.12) with $\zeta = 0$ is of Fredholm type, the existence of its solution will be proved if one proves that equation (2.15) with $\zeta = 0$ has only the trivial solution. Let us prove this. The equation

$$\frac{A\sigma - \sigma}{2} = 0 \tag{2.56}$$

implies that $u_N^- = 0$ on S. By theorem 2.1.7 it follows that $u = 0$ in D'. Therefore, $u = 0$ on S and u solves the Dirichlet problem for equation (2.1) in D. This implies that $u = 0$ in D because k^2 cannot be a Dirichlet eigenvalue of the Laplacian if D is sufficiently small. If $u = 0$ in $D \cup D'$ then $\sigma = u_N^+ - u_N^- = 0$. By the Fredholm alternative the existence of the solution to equation (2.12) with $\zeta = 0$ is proved.

Theorem 2.1.8 is proved. $\qquad\square$

Let us now prove an analog of theorem 2.1.3.

Formula (2.24) is now taking the form

$$Q_1 := \int_S e^{-ik\beta \cdot t}\sigma(t)dt \sim \int_S \sigma(t)dt - ik\beta_p \int_S t_p\sigma(t)dt, \tag{2.57}$$

where *here and below over repeated indices summation is understood.*

The novel point in a study of the scattering problem with the Neumann boundary condition is the necessity to use both terms (2.57) because they are of the same order as $a \to 0$, namely, they are both $O(a^3)$. This is in contrast with the problem with the Dirichlet boundary condition where $Q = O(a)$, and with the problem with impedance boundary condition where $Q = O(\zeta a^2)$ as $a \to 0$. Moreover, we prove that the scattering in the problem with Neumann boundary condition is *anisotropic*, in sharp contrast with the cases of the Dirichlet and impedance boundary conditions.

Let us first estimate

$$Q = \int_S \sigma(t)dt.$$

We look for the solution to problem (2.1), (2.55), (2.3), (2.4) of the form (2.10). The boundary condition (2.55) yields the integral equation for σ:

$$\sigma = A\sigma + 2u_{0N} \quad \text{on } S. \tag{2.58}$$

Integrate this equation over S and use formula (2.30) to get

$$Q \sim \int_S u_{0N}ds = \int_D \nabla^2 u_0 dx = \nabla^2 u_0(x_1)|D|, \tag{2.59}$$

where $x_1 \in D$ is an arbitrary point inside D and we took into account that because D is small $\nabla^2 u_0(x_1)$ does not depend on the choice of x_1. Since $D = O(a^3)$, it follows from equation (2.59) that $Q = O(a^3)$ as was mentioned earlier.

Let us estimate the last integral in equation (2.57). Introduce the matrix (tensor) β_{pq} by the formula

$$\beta_{pq} := \frac{1}{|D|}\int_S t_p\sigma_q(t)dt, \tag{2.60}$$

where σ_q is the unique solution to the equation

$$\sigma_q = A\sigma_q - 2N_q, \quad q = 1, 2, 3, \tag{2.61}$$

and $N_q = N \cdot e_q$, where $\{e_q\}$ is an orthonormal Cartesian basis of $\mathbb{R}^3$. equation (2.61) is of Fredholm type and it has a solution because the corresponding homogeneous equation $\sigma = A\sigma$ has only the trivial solution $\sigma = 0$. Indeed, this equation is equivalent to the relation $u_N^- = 0$, and $u = \int_S g(x, t)\sigma(t)dt$ satisfies equation (2.1) and the radiation condition (2.4). Therefore $u = 0$ in D'. Consequently $u = 0$ on S. Thus, u solves equation (2.1) in D and vanishes on S. This means that k^2 is the Dirichlet eigenvalue of the Laplacian in D, which is a contradiction because D is assumed to be sufficiently small. Therefore $u = 0$ in $D \cup D'$, and $\sigma = u_N^+ - u_N^- = 0$. This proves the existence and uniqueness of the solution σ_q to equation (2.61). The function σ in equation (2.57) solves equation (2.58), where

$$u_{0N} = \frac{\partial u_0}{\partial x_q}N_q. \tag{2.62}$$

Recall that summation over q is understood. Thus

$$2u_{0N} = -2N_q\left(-\frac{\partial u_0(x_1)}{\partial x_q}\right). \tag{2.63}$$

From equations (2.58), (2.63), (2.61), and (2.60) one gets

$$-ik\beta_p \int_S t_p\sigma(t)dt = ik\beta_{pq}\beta_p\frac{\partial u_0(x_1)}{\partial x_q}|D|, \tag{2.64}$$

where $|D|$ is the volume of D, β_{pq} is the tensor defined in (2.60), and β_p is the pth component of the unit vector $\beta := \frac{x-x_1}{|x-x_1|}$, $x_1 \in D$.

Therefore,

$$Q_1 \sim |D|\left(\nabla^2 u_0(x_1) + ik\beta_{pq}\beta_p\frac{\partial u_0(x_1)}{\partial x_q}\right), \quad x_1 \in D, \tag{2.65}$$

where Q_1 is defined in (2.57).

Consequently, an analog of formulas (2.44) and (2.45) in the case of the Neumann boundary condition takes the form

$$u(x) = u_0(x) + g(x, x_1)\left(\nabla^2 u_0(x_1) + ik\beta_{pq}\beta_p\frac{\partial u_0(x_1)}{\partial x_q}\right)|D|, \quad |x - x_1| \gg a, \tag{2.66}$$

where

$$\beta = \frac{x - x_1}{|x - x_1|}, \quad x_1 \in D, \tag{2.67}$$

and

$$A(\beta, \alpha, k) = \left(\nabla^2 u_0(x_1) + ik\beta_{pq}\beta_p \frac{\partial u_0(x_1)}{\partial x_q}\right)\frac{|D|}{4\pi}. \qquad (2.68)$$

If x_1 is the origin, $x_1 = 0$, and $u_0 = e^{ik\alpha \cdot x}$, then formula (2.68) can be rewritten as

$$A(\beta, \alpha, k) = -\frac{k^2|D|}{4\pi}\left(1 + \beta_{pq}\beta_p\alpha_q\right). \qquad (2.69)$$

One can see from this formula that $A = O(a^3)$, the scattering is anisotropic and its anisotropic part is described by the tensor β_{pq} defined by formula (2.60).

Let us summarize the results.

Theorem 2.1.9. *The scattering problem (2.1), (2.55), (2.3), (2.4) is uniquely solvable. Their solutions can be calculated by formulas (2.66), (2.67).*

The scattering amplitude $A(\beta, \alpha, k)$ is given by formula (2.68), $A(\beta, \alpha, k) = O(a^3)$, and the scattering is anisotropic.

2.1.4 The interface (transmission) boundary condition

Consider the following scattering problem

$$(\nabla^2 + k^2)u = 0 \quad \text{in } D', \quad k^2 > 0, \qquad (2.70)$$

$$(\nabla^2 + k_1^2)u = 0 \quad \text{in } D, \quad k_1^2 > 0, \qquad (2.71)$$

$$u^+ = u^-, \quad \rho_1 u_N^+ = u_N^-, \quad \rho_1 > 0, \qquad (2.72)$$

$$u = u_0 + v, \qquad (2.73)$$

$$\frac{\partial v}{\partial r} - ikv = o\left(\frac{1}{r}\right), \quad r = |x| \to \infty, \qquad (2.74)$$

where u_0 is the incident field,

$$(\nabla^2 + k^2)u_0 = 0 \quad \text{in } \mathbb{R}^3. \qquad (2.75)$$

In particular, the plane wave incident field $u_0 = e^{ik\alpha \cdot x}$, $\alpha \in S^2$, is often considered. The number ρ_1 in the transmission boundary condition (2.72) is assumed positive and $\rho_1 \neq 1$. The number k_1^2 in equation (2.71) is not equal to k^2.

Physically problem (2.70)–(2.75) corresponds to the incident wave propagating through D and in this process the wave is scattered.

As earlier, the first task is to prove the existence and uniqueness of the solution to the scattering problem (2.70)–(2.75).

Theorem 2.1.10. *The above problem with* $\rho_1 = \text{const} > 0$ *has no more than one solution.*

Proof. It is sufficient to prove that the homogeneous problem, that is, the problem with $u_0 = 0$, has only the trivial solution. To prove this, multiply equation (2.70) and (2.71) by $\bar{u}$, the bar stands for complex conjugate, integrate the second equation over D, the first equation over $D_R' := B_R \cap D'$, $B_R := \{x : |x| \leqslant R\}$, then use Green's formula and condition (2.72) and (2.74) and get

$$
\int_S (\bar{u} u_N^+ - u \overline{u_N^+}) ds - \int_S (\bar{u} u_N^- - u \overline{u_N^-}) ds + \int_{S_R} (\bar{u} u_N - u \overline{u_N}) ds
$$
$$
= \int_S [\bar{u} u_N^+ (1 - \rho_1) - u \overline{u_N^+}(1 - \rho_1)] ds + 2ik \int_{S_R} |u|^2 ds + o(1) \qquad (2.76)
$$
$$
= 0,
$$

where $\lim_{R \to \infty} o(1) = 0$.

One has, using Green's formula,

$$
(1 - \rho_1) \int_S (\bar{u} u_N^+ - u \overline{u_N^+}) ds = 0, \qquad (2.77)
$$

because u and $\bar{u}$ solve the same equation (2.71).

Thus, relation (2.76) implies

$$
\lim_{R \to \infty} \int_{S_R} |u|^2 ds = 0. \qquad (2.78)
$$

From this, equation (2.70) and the radiation condition (2.74) with $u_0 = 0$, $u = v$, it follows that $u = 0$ in D'. Thus, $u^- = u_N^- = 0$.

Consequently, $u^+ = u_N^+ = 0$, and $u = 0$ in D.

Theorem 2.1.10 is proved. $\qquad\qquad\qquad\qquad\qquad\qquad\qquad\qquad\qquad\qquad\quad\square$

Remark 2.1.3. *The proof remains valid with only a slight change if one assumes that* $\text{Im}\, k^2 \geqslant 0$.

Theorem 2.1.11. *Problem (2.70)–(2.75) has a solution.*

Proof. Let us look for a solution of the form

$$
u(x) = u_0(x) + \int_S g(x, t)\sigma(t)dt + \kappa \int_D g(x, y)u(y)dy, \qquad (2.79)
$$

where

$$
\kappa := k_1^2 - k^2, \quad g(x, y) = \frac{e^{ik|x-y|}}{4\pi|x - y|}, \qquad (2.80)
$$

and σ has to be found such that conditions (2.72) are satisfied.

Note that equations (2.70), (2.71), (2.73), (2.74), and (2.75) are satisfied.

This is obvious for equations (2.73)–(2.75) and follows for equations (2.70), (2.71) from the equation

$$(\nabla^2 + k^2)g(x, y) = -\delta(x - y). \tag{2.81}$$

The first condition (2.72) is satisfied because both integrals in equation (2.79) are continuous functions in $\mathbb{R}^3$.

The second condition (2.72) is satisfied if σ and u solve the system of integral equations (2.79) and (2.82), where

$$\rho_1\frac{A\sigma + \sigma}{2} - \frac{A\sigma - \sigma}{2} + (\rho_1 - 1)\frac{\partial}{\partial N_S}Bu + (\rho - 1)u_{0N} = 0. \tag{2.82}$$

Here

$$A\sigma = 2\int_S \frac{\partial g(s, t)}{\partial N_S}\sigma(t)dt, \quad Bu = \kappa\int_D g(x, y)u(y)dy. \tag{2.83}$$

Let us rewrite equation (2.82) as follows:

$$\sigma = \lambda A\sigma + 2\lambda B_1 u + 2\lambda u_{0N}, \tag{2.84}$$

where

$$\lambda = \frac{1 - \rho}{1 + \rho}, \quad B_1 u := \frac{\partial(Bu)}{\partial N_S}. \tag{2.85}$$

If $0 < \rho < \infty$ then $\lambda \in (-1, 1)$. We assume, as earlier, that

$$ka \ll 1. \tag{2.86}$$

If the pair $\{\sigma, u(x)|_{x\in D}\}$ is found from the system of integral equations (2.79) and (2.84), then the solution to the scattering problem (2.70)–(2.75) is found in $\mathbb{R}^3$ by formula (2.79). □

Theorem 2.1.12. *The system (2.79)–(2.84) has a unique solution. Thus, the scattering problem (2.70)–(2.75) has a unique solution.*

Proof. Equations (2.79)–(2.84) are of Fredholm type. Therefore it is sufficient to prove that the system of homogeneous versions of these equations has only the trivial solution.

From the derivation of equations (2.79) and (2.84) it follows that the homogeneous versions of these equations imply that u solves the scattering problem (2.70)–(2.75) with $u_0 = 0$.

By theorem 2.1.10 this problem has only the trivial solution $u = 0$. Therefore the system of equations (2.79) and (2.84) has a solution and this solution is unique, and the scattering problem (2.70)–(2.75) has a solution and this solution is unique.

Theorems 2.1.11 and 2.1.12 are proved. □

Let us now find analytical formulas for the solution of the scattering problem (2.70)–(2.75). It follows from formula (2.79) that

$$u(x) \sim u_0(x) + g(x, x_1)\left[\int_S e^{-ik\beta \cdot t}\sigma(t)dt + \kappa u(x_1)|D|\right],$$

$$|x - x_1| \gg a, \quad x_1 \in D. \tag{2.87}$$

To make this formula practically applicable one needs to calculate the main terms of the asymptotic of the surface integral in equation (2.87) and of the term $u(x_1)$.

Note that

$$g(s, t) = g_0(s, t)[1 + O(ka)], \quad g_0(s, t) = \frac{1}{4\pi|s - t|}, \quad s, t \in S, \tag{2.88}$$

$$\frac{\partial g(s, t)}{\partial N_S} = \frac{\partial g_0}{\partial N_S}[1 + O(k^2 a^2)], \quad a \to 0, \tag{2.89}$$

$$A\sigma = 2\int_S \frac{\partial g(s, t)}{\partial N_S}\sigma(t)dt \sim 2\int_S \frac{\partial g_0(s, t)}{\partial N_S}\sigma(t)dt = A_0\sigma, \tag{2.90}$$

$$Bu \sim \kappa\int_B g_0(x, y)u(y)dy = \kappa B_0 u, \tag{2.91}$$

$$B_1 u \sim B_{01} u, \quad B_{01} u = \kappa\int_B \frac{\partial}{\partial N_S}g(s, y)u(y)dy. \tag{2.92}$$

One has

$$\int_S e^{-ik\beta \cdot t}\sigma(t)dt \sim \int_S \sigma(t)dt - ik\beta_p\int_S t_p\sigma(t)dt. \tag{2.93}$$

Let

$$Q := \int_S \sigma(t)dt, \quad Q_1 := Q - ik\beta_p\int_S t_p\sigma(t)dt. \tag{2.94}$$

Then, by formula (2.87), one gets

$$u(x) \sim u_0(x) + g(x, x_1)[Q_1 + \kappa u(x_1)|D|], \quad |x - x_1| \gg a. \tag{2.95}$$

Let us derive analytical formula for Q_1 and $u(x_1)$. Integrate equation (2.84) over S and get

$$Q := \lambda\int_S A\sigma ds + 2\lambda\int_S B_1 u ds + 2\lambda\int_S u_{0N}ds. \tag{2.96}$$

One has, by the divergence theorem,

$$2\lambda\int_S u_{0N}ds \sim 2\lambda\nabla^2 u_0(x_1)|D|. \tag{2.97}$$

Furthermore, using formula (2.34), one gets:

$$\lambda \int_S A\sigma ds \sim \lambda \int_S A_0 \sigma ds = -\lambda \int_S \sigma ds = -\lambda Q, \qquad (2.98)$$

and

$$2\lambda \int_S B_1 u ds = 2\lambda\kappa \int_D dx \nabla_x^2 \int_D g(x,y)u(y)dy.$$

Equation (2.81) implies

$$\int_D dx \nabla_x^2 \int_D g(x,y)u(y)dy = -k^2 \int_D dx \int_D g(x,y)u(y)dy - \int_D dx u(x) \\ \sim -u(x_1)|D|. \qquad (2.99)$$

Thus,

$$2\lambda \int_S B_1 u ds \sim -2\lambda\kappa u(x_1)|D|, \quad a \to 0. \qquad (2.100)$$

Therefore, equations (2.96)–(2.100) imply

$$Q \sim -\frac{2\lambda\kappa}{1+\lambda}u(x_1)|D| + \frac{2\lambda}{1+\lambda}\nabla^2 u_0(x_1)|D|. \qquad (2.101)$$

In order to estimate $u(x_1)$ as $a \to 0$ let us integrate equation (2.79) over D and obtain

$$u(x_1)|D| \sim u_0(x_1)|D| + \int_D dx \int_S g(x,t)\sigma(t)dt + \kappa \int_D dx \int_D g(x,y)u(y)dy. \quad (2.102)$$

One has

$$\int_D dx \int_S g(x,t)\sigma(t)dt = \int_S dt\sigma(t) \int_D dx g(x,t) \sim QO(a^2), \qquad (2.103)$$

and

$$\int_D dx \int_D g(x,y)u(y)dy = u(x_1)|D|O(a^2), \qquad (2.104)$$

where we have used the relation

$$\int_D g(x,y)dy = O(a^2), \quad a \to 0, \qquad (2.105)$$

which holds if 0.5 diam $D \leqslant a$.

It follows from formulas (2.102)–(2.105) that

$$u(x_1)|D| \sim u_0(x_1)|D| + QO(a^2), \quad a \to 0, \qquad (2.106)$$

where we have neglected the terms of higher order of smallness as $a \to 0$.

From equations (2.101) and (2.106) it follows that

$$Q \sim \frac{2\lambda}{1+\lambda}|D|(-\kappa u_0(x_1) + \nabla^2 u_0(x_1)), \quad a \to 0, \tag{2.107}$$

and

$$u(x_1) \sim u_0(x_1), \quad a \to 0 \tag{2.108}$$

Let us derive an analytical formula for the second integral in equation (2.94).

Multiply equation (2.84) by t_p and integrate over S. Take into account relation (2.98) and formulas (2.100) and (2.101) to get

$$2\lambda \int_S ds s_p B_1 u \sim O(a^4), \quad a \to 0. \tag{2.109}$$

Let us define now an analog of the matrix (2.60):

$$\beta_{pq}(\lambda) := \frac{1}{|D|} \int_S t_p \sigma_q(t) dt, \tag{2.110}$$

where the function $\sigma_q(t) := \sigma_q(t, \lambda)$ solves the equation

$$\sigma_q(t) = \lambda A \sigma_q(t) - 2\lambda N_q. \tag{2.111}$$

Since $|\lambda| = |\frac{1-\rho}{1+\rho}| < 1$ when $\rho > 0$, and the operator A has no eigenvalues in the interval $(-1, 1)$, equation (2.111) has a unique solution.

One has $u_{0N}(t) \sim u_{0N}(x_1)$, $x_1 \in D$, and

$$2\lambda u_{0N} = -2\lambda N_q \left(-\frac{\partial u_0(x_1)}{\partial x_q} \right), \quad \sigma \sim -\sigma_q \frac{\partial u_0(x_1)}{\partial x_q}. \tag{2.112}$$

Therefore, neglecting the term (2.109), which is of higher order of smallness as $a \to 0$, one gets

$$\int_S t_p \sigma(t) dt = -|D|\beta_{pq}(\lambda) \frac{\partial u_0(x_1)}{\partial x_q}. \tag{2.113}$$

Consequently,

$$-ik\beta_p \int_S t_p \sigma(t) dt = ik\beta_{pq}(\lambda)\beta_p \frac{\partial u_0(x_1)}{\partial x_q}|D|, \tag{2.114}$$

and formulas (2.94), (2.107), and (2.114) yield the formula for Q_1:

$$Q_1 = \frac{2\lambda}{1+\lambda}|D|(\nabla^2 u_0(x_1) - \kappa u_0(x_1)) + ik\beta_{pq}(\lambda)\beta_p \frac{\partial u_0(x_1)}{\partial x_q}|D| \tag{2.115}$$

where

$$\frac{2\lambda}{1+\lambda} = 1 - \rho, \quad \beta_p := \frac{(x - x_1)_p}{|x - x_1|}, \quad x_p := x \cdot e_p. \tag{2.116}$$

From formulas (2.87), (2.95), (2.108), and (2.115) it follows that

$$u(x) \sim u_0(x) + g(x, x_1)\left[(1 - \rho)(\nabla^2 u_0(x_1) - \kappa u_0(x_1)) + ik\beta_{pq}(\lambda)\frac{(x - x_1)_p}{|x - x_1|} \right.$$
$$\left. \frac{\partial u_0(x_1)}{\partial x_q} + \kappa u_0(x_1)\right]|D|, \quad |x - x_1| \gg a. \tag{2.117}$$

Furthermore,

$$A(\beta, \alpha, k) = \frac{|D|}{4\pi}\left[(1 - \rho)(\nabla^2 u_0(x_1) - \kappa u_0(x_1)) + ik\beta_{pq}(\lambda)\beta_p\frac{\partial u_0(x_1)}{\partial x_q} + \kappa u_0(x_1)\right]. \tag{2.118}$$

Formulas (2.117) and (2.118) give our final result.
Note that $|A(\beta, \alpha, k)| = O(a^3)$ and the scattering is anisotropic.
Let us summarize the results.

Theorem 2.1.13. *If $ka \ll 1$ and $\rho > 0$, then the scattering problem (2.70)–(2.75) has a unique solution. This solution can be calculated by formula (2.117). The scattering amplitude is calculated by formula (2.118).*

2.1.5 Summary of the results

The results of this chapter can be summarized as follows.

Scattering problem (2.1)–(2.4) has a unique solution for any ζ, $\operatorname{Im}\zeta \leqslant 0$, including the limiting cases $\zeta = 0$ and $\zeta = \infty$.

The solution can be calculated by formula (2.44) if $ka \ll 1$ and the scattering amplitude by formula (2.46).

One has $|A(\beta, \alpha, k)| = O(\zeta a^2)$ and the scattering is isotropic. If $\zeta = \infty$, then u can be calculated by formula (2.52) and the scattering amplitude-by formula (2.53) for $ka \ll 1$. The scattering is isotropic and $|A(\beta, \alpha, k)| = O(a)$.

If $\zeta = 0$, then u can be calculated by formula (2.66), the scattering amplitude-by formula (2.68) if $ka \ll 1$, $|A(\beta, \alpha, k)| = O(a^3)$, and the scattering is anisotropic.

The scattering problem (2.70)–(2.75) with the interface (transmission) boundary condition has a unique solution if $\rho \geqslant 0$, $k^2 > 0$, $k_1^2 > 0$. If $ka \ll 1$ then this solution can be calculated by formula (2.117), the scattering amplitude-by formula (2.118), $|A(\beta, \alpha, k)| = O(a^3)$, and the scattering is anisotropic.

2.2 Scalar wave scattering by many small bodies of an arbitrary shape

2.2.1 Impedance bodies

Consider the many-body scattering problem in the case of small impedance bodies (particle) of an arbitrary shape:

$$(\nabla^2 + k^2)u = 0 \quad \text{in } D', \quad D' = R^3 \backslash D, \tag{2.119}$$

where

$$D = \bigcup_{m=1}^{M} D_m, \quad a = \max_m\left(\frac{1}{2}\text{diam } D_m\right), \quad ka \ll 1,$$

$$\left.\frac{\partial u}{\partial N}\right|_{S_m} = \zeta_m u, \quad 1 \leqslant m \leqslant M, \quad \text{Im } \zeta_m \leqslant 0, \tag{2.120}$$

$$u = u_0 + v, \tag{2.121}$$

$$\frac{\partial v}{\partial r} - ikv = o\left(\frac{1}{r}\right), \quad r = |x| \to \infty, \quad \frac{x}{|x|} := \beta := x^0. \tag{2.122}$$

The field v is the scattered field, u_0 is the incident field which satisfies equation (2.119) in $\mathbb{R}^3$. For example, the incident field is often the direction of the plane wave $u_0 = e^{ika \cdot x}$, $\alpha \in S^2$ describes the direction of the propagation of the incident wave, $k > 0$ is the wave number, N is the unit normal to $S := \bigcup_{m=1}^{M} S_m$ pointing into D', $S_m \in C^{1,2}$, $\lambda \in (0, 1]$.

Theorem 2.2.1. *Problem (2.119)–(2.122) has a unique solution.*

Proof. The proof is essentially the same as the proof of theorem 2.1.1, and we leave the details to the reader.
Theorem 2.2.1 is proved. $\qquad\square$

Let us look for the solution to problem (2.119)–(2.122) of the form

$$u(x) = u_0(x) + \sum_{m=1}^{M} \int_{S_m} g(x, t)\sigma_m(t)dt. \tag{2.123}$$

This function solves problem (2.119)–(2.122) if and only if σ_m are chosen so that the boundary condition (2.120) is satisfied. These boundary conditions lead to a system of M integral equations. Since M can be very large, say, $M = 10^{12}$, it is not practically feasible to solve such a large system of simultaneous integral equations for unknown function σ_m. By this reason we develop a new approach to solving problem (2.119)–(2.122). This approach uses essentially the assumption $ka \ll 1$. It consists in replacing the unfeasible task of finding M unknown numbers. This task is practically feasible and, moreover, it is justified physically.

Let us rewrite equation (2.123) as

$$u(x) = u_0(x) + \sum_{m=1}^{M} g(x, x_m)Q_m + \sum_{m=1}^{M} \int_{S_m} [g(x, t) - g(x, x_m)]\sigma_m(t)dt, \quad (2.124)$$

where

$$Q_m := \int_{S_m} \sigma_m(t)dt, \quad x_m \in D_m. \quad (2.125)$$

The numbers Q_m are the numbers we mentioned above. The central idea is to show that the second sum in equation (2.124) is negligible compared with the first as $a \to 0$. If this is done, then the solution to the scattering problem (2.119)–(2.122) is given by the formula

$$u(x) \sim u_0(x) + \sum_{m=1}^{M} g(x, x_m)Q_m, \quad (2.126)$$

so that the scattering problem is solved if the numbers Q_m are found.

These numbers will be found from a linear algebra system. To derive this system let us introduce the notion of the effective field acting on the jth small body. We denote this field $u_e(x)$ and define it as

$$u_e(x) := u_0(x) + \sum_{m \neq j}^{M} g(x, x_m)Q_m. \quad (2.127)$$

Let us make the basic assumptions that hold throughout the book for the many-body wave scattering in the case of the small bodies.

Assumption A. The minimal distance d between neighboring bodies is much larger than the size a of the bodies:

$$d \gg a. \quad (2.128)$$

The number $\mathcal{N}(\Delta)$ of the small bodies in an arbitrary open set Δ is given by the formula:

$$\mathcal{N}(\Delta) = \frac{1}{a^{2-\kappa}} \int_{\Delta} N(x)dx[1 + o(1)], \quad a \to 0, \quad (2.129)$$

where $N(x) \geqslant 0$ is a continuous function, $\kappa \in [0, 1)$ is a number,

$$\zeta_m = \frac{h(x_m)}{a^\kappa}, \quad \text{Im } h(x) \leqslant 0, \quad (2.130)$$

and $h(x)$ is a continuous function, $1 \leqslant m \leqslant M$.

The parameter κ, the function h, and the function $N(x) \geqslant 0$ can be chosen by the experimentalist as he (she) wishes.

The assumption Im $h \leqslant 0$ guarantees by theorem 2.2.1 the uniqueness of the solution to problem (2.119)–(2.122). The impedance function ζ does not have to depend on a as in (2.130), but no physical restrictions prevent such a dependence. We will see in chapter 3 that such a dependence can be used practically. We assume that the small bodies are distributed according to the law (2.129) in an arbitrary finite region Ω, so that $N(x) = 0$ in $\Omega' = \mathbb{R}^3 \backslash \Omega$.

Assumption (2.128) allows one to consider jth body as the body placed in the exterior field $u_e(x)$ defined by formula (2.127). We do not put index j on $u_e(x) := u_{e,j}(x)$ to keep the notations simpler.

Let us denote

$$u_e(x_j) := u_j, \quad 1 \leqslant j \leqslant M. \tag{2.131}$$

Our first task is to express Q_m in terms of u_m. This is done by formula (2.42) in which $u_0(x)$ is replaced by $u_e(x)$:

$$Q_m = -\zeta_m |S_m| u_{e,m}(x_m) := -\zeta_m |S_m| u_m, \tag{2.132}$$

and we use the notation $u_m := u_{e,m}$ for simplicity.

If

$$S_m = c_m a^2, \quad c_m = c = \text{const},$$

then assumption (2.130) and formula (2.132) yield

$$Q_m = -c_m h_m a^{2-\kappa} u_m, \quad 1 \leqslant m \leqslant M, \quad h_m := h(x_m). \tag{2.133}$$

From equations (2.133) and (2.127) it follows that

$$u_j = u_{0j} - \sum_{\substack{m=1 \\ m \neq j}}^{M} g_{jm} c_m h_m a^{2-\kappa} u_m, \quad 1 \leqslant j \leqslant M, \tag{2.134}$$

where

$$g_{jm} := g(x_j, x_m), \quad u_{0j} := u_0(x_j). \tag{2.135}$$

Linear algebraic system (2.134) (= LAS1) allows one to find numbers u_m, $1 \leqslant m \leqslant M$. If these numbers are found, then the numbers Q_m are found by formula (2.133) and the solution to the scattering problem (2.119)–(2.122) is found by formula (2.126).

Let us prove that the neglected term in formula (2.124) is much smaller than the term we kept.

Lemma 2.2.1. One has

$$J_1 := |g(x, x_m) Q_m| \gg \left| \int_{S_m} [g(x, t) - g(x, x_m)] \sigma_m(t) dt \right| := J_2, \quad a \to 0. \tag{2.136}$$

Proof. By equation (2.128) one has

$$|g(x, x_m)Q_m| \geqslant \frac{Q_m}{4\pi d}, \qquad |x - x_m| \geqslant d. \tag{2.137}$$

Since $|x - x_m| \geqslant d \gg a$ and $|t - x_m| \leqslant a$, one has

$$|g(x, t) - g(x, x_m)| \leqslant \max\left\{ O\left(\frac{a}{d^2}\right), O\left(\frac{ka}{d}\right) \right\}, \tag{2.138}$$

and

$$\frac{J_2}{J_1} \leqslant O\left(\frac{a}{d} + ka\right) \ll 1. \tag{2.139}$$

Lemma 2.2.1 is proved. $\qquad\qquad\qquad\qquad\qquad\qquad\qquad\qquad\qquad\qquad\qquad\square$

Our next task is to reduce the order of the linear algebraic system (2.134). For practical calculations this is an important task. Let us explain how to do this.

Let us partition the domain Ω, where the small bodies are distributed, into a union of P cubes Δ_p. These cubes do not have common interior points but may have common parts of the boundary. Let $x_p \in \Delta_p$. We assume that the side of Δ_p is b and the following conditions hold:

$$a \ll d \ll b. \tag{2.140}$$

One assumes that $d = d(a)$ and $b = b(a)$. Then, the assumption, which implies equation (2.140), can be written as

$$\lim_{a\to 0} \frac{a}{d} = 0, \quad \lim_{a\to 0} \frac{d}{b} = 0. \tag{2.141}$$

One can rewrite equation (2.134) as

$$u_q = u_{0q} - \sum_{p \neq q}^{P} c_p g_{qp} h_p u_p a^{2-\kappa} \sum_{x_m \in \Delta_p} 1, \quad 1 \leqslant q \leqslant P. \tag{2.142}$$

If $\operatorname{diam} \Delta_p \ll 1$ and $N(x)$ is continuous, then

$$a^{2-\kappa} \mathcal{N}(\Delta_p) = \sum_{x_m \in \Delta_p} 1 = N(x_p)|\Delta_p|[1 + o(1)], \quad a \to 0. \tag{2.143}$$

It follows from equations (2.142) and (2.143) that, assuming $c_p = c = \text{const}$, $|\Delta_p| = b^3$ is the volume of Δ_p, one has

$$u_q = u_{0q} - c \sum_{p \neq q}^{P} g_{qp} h_p N_p u_p |\Delta_p|, \quad 1 \leqslant q \leqslant P, \tag{2.144}$$

where the $o(1)$ term in equation (2.143) is neglected and the continuity of the functions $h(x)$, $N(x)$ and $u(x)$ allows one to replace h_m and u_m by h_p and u_p for any $x_m \in \Delta_p$, and to use the formula

$$\int_{\Delta_p} N(x)dx \sim N(x_p)|\Delta_p|, \quad a \to 0.$$

Equation (2.144) is the linear algebraic system (LAS2) of the order $P \ll M$.

This concludes the description of the process of the reduction of the order of LAS (2.16).

One can see that LAS (2.144) is obtained if one uses the collocation method (see [2]) for solving the integral equation

$$u(x) = u_0(x) - c \int_\Omega g(x, y)h(y)N(y)u(y)dy. \tag{2.145}$$

Integral equation (2.145) is the equation for the limiting effective field in the medium in which many small particles are embedded when the size of a particle tends to zero, $a \to 0$, while the number of the particles $\mathcal{N}(D) \to \infty$ according to the law (2.129), that is $\mathcal{N}(D) = O(\frac{1}{a^{2-\kappa}})$.

Lemma 2.2.2. If $q = ch(x)N(x)$, $\mathrm{Im}\, q \leqslant 0$ and q is compactly supported, then equation (2.145) has a unique solution.

Proof. Indeed, it is a Fredholm-type equation, that is, an equation for which the Fredholm alternative is valid (see [2]). Moreover, the homogeneous version

$$u(x) = -c \int_\Omega g(x, y)h(y)N(y)u(y)dy$$

of this equation has only the trivial solution. To prove this, apply the operator $\nabla^2 + k^2$ to the above equation and get

$$[\nabla^2 + k^2 - cN(x)h(x)]u = 0 \quad \text{in } \mathbb{R}^3. \tag{2.146}$$

This is a Schrödinger equation with a compactly supported potential $q(x) = cN(x)h(x)$, $q(x) = 0$ in Ω'. If $\mathrm{Im}\, h \leqslant 0$ then the solution to equation (2.146) satisfying the radiation condition (2.122) must be identically equal to zero. To prove this, multiply equation (2.146) by $\bar{u}$, subtract the complex conjugate of equation (2.146) multiplied by u, integrate over the ball $B_R := \{x : |x| \leqslant R\}$ of large radius R and use the radiation condition. The result is

$$\lim_{R \to \infty} \left(2ik \int_{|s|=R} |u|^2 ds - 2i \int_{B_R} \mathrm{Im}\, q(x)|u|^2 dx \right) = 0. \tag{2.147}$$

If $\text{Im } q \leqslant 0$, then one can concludes that

$$\lim_{R \to \infty} \int_{|s|=R} |u|^2 ds = 0. \qquad (2.148)$$

This and equation (2.146) imply that $u = 0$ in Ω', and, by the unique continuation theorem for solutions to elliptic equation (2.146), it follows that $u = 0$ in $\mathbb{R}^3$. Lemma 2.2.2 is proved. $\qquad \square$

Consequently, by the Fredholm alternative, the operator $(I + T)^{-1}$ is bounded, where

$$Tu := c \int_\Omega g(x, y)h(y)N(y)u(y)dy.$$

Remark 2.2.1. *Let us denote $\eta := \text{Im } h(x)$. The operator $T = T_\eta$ depends continuously on the parameter η in the norm of the operators in the space $C(\Omega)$. Since the operator $(I + T_\eta)^{-1}$ is bounded for $\text{Im } \eta \leqslant 0$ and depends on η continuously in the norm of the operators, it must be bounded for sufficiently small positive $\eta > 0$. For such η one has $\arg(-cN(x)h(x)) < 0$.*

Convergence of the collocation method for solving integral equation (2.145) is studied in detail in [2], where a one-to-one correspondence between the function that solves equation (2.145) approximately with any desired accuracy and the vector $(u_1, \dots, u_P)$ that solves LAS (2.144) is established. This convergence justifies the transition in the limit $a \to 0$ from the LAS (2.144) to the integral equation (2.145). In this sense it plays the role of the homogenization theory (see [4]), but our theory does not require the periodicity assumption, used in the theory, the spectrum of our problem (2.119)–(2.122) is continuous and not discrete, as in the usual theory, and our operator is non-selfadjoint, also in contrast to the standard theory.

The dependence of c_p on p in formula (2.142) can describe the particles which are not identical. For simplicity we have assumed in equation (2.144) that $c_p = c$ does not depend on p, that is, that all the particles are identical.

Let us apply the operator $\nabla^2 + k^2$ to equation (2.145) and use the known equation

$$(\nabla^2 + k^2)g(x, y) = -\delta(x - y)$$

to get equation (2.146). This equation can be interpreted physically as follows: the limiting medium, obtained by embedding many small particles according to the distribution law (2.129) with the boundary impedances defined in equation (2.130), has new refraction coefficient defined by the relation

$$n^2(x) = 1 - ck^{-2}h(x)N(x), \qquad (2.149)$$

so

$$n(x) = [1 - ck^{-2}h(x)N(x)]^{1/2}. \tag{2.150}$$

Since the functions $h(x)$, Im $h(x) \leqslant 0$, and $N(x) \geqslant 0$, can be chosen by an experimentalist formula, (2.150) gives very wide possibilities for creating materials with a desired refraction coefficient by embedding into a given material many small impedance particles according to the distribution law (2.129) with boundary impedances ζ_m defined in equation (2.130).

This is discussed in chapter 3.

Let us summarize the results.

Theorem 2.2.2. *Problem (2.119)–(2.122) has a unique solution which can be calculated by formula (2.126) in which the numbers Q_m are given by formula (2.132) and the numbers u_m are found from the linear algebraic system (2.134). As $a \to 0$ and assumptions (2.128)–(2.130), (2.140) hold, then the field tends to the (unique) solution of the integral equation (2.145). The resulting limiting medium has the refraction coefficient defined in equation (2.150).*

2.2.2 The Dirichlet boundary condition

Let us consider the scattering problem (2.119)–(2.122) with the impedance boundary condition (2.120) replaced by the Dirichlet boundary condition

$$u|_{S_m} = 0, \quad 1 \leqslant m \leqslant M. \tag{2.151}$$

Theorem 2.2.3. *Problem (2.119), (2.151), (2.121), (2.122) has a unique solution.*

Proof. Let us first prove that the above problem has no more than one solution. This is equivalent to proving that the corresponding homogeneous problem, that is the problem with $u_0 = 0$, has only the trivial solution. Let u be a solution to homogeneous problem (2.119), (2.151), (2.121), (2.122). Multiply equation (2.119) by $\bar{u}$, integrate over $D_R' := D' \cap B_R$ and use Green's formula to get

$$0 = \int_{D_R'} (k^2|u|^2)dx + ik \int_{S_R'} |u|^2 ds + o(1), \quad R \to \infty. \tag{2.152}$$

Taking the imaginary part of equation (2.152) one obtains

$$\lim_{R \to \infty} \int_{S_R} |u|^2 ds = 0. \tag{2.153}$$

This and the equation (2.119) imply that $u = 0$ in D'. Thus, it is proved that problem (2.119), (2.151), (2.121), (2.122) has no more than one solution.

Let us prove the existence of the solution of the form (2.123). It is sufficient to prove that σ_m are uniquely defined by the boundary condition (2.151):

$$\sum_{m=1}^{M} \int_{S_m'} g(s, t)\sigma_m(t)dt = -u_0(s), \quad s \in S_j, \quad 1 \leqslant j \leqslant M. \tag{2.154}$$

Let us rewrite this equation as

$$\sigma_j + \sum_{m=1}^{M} T_{jj}^{-1}T_{jm}\sigma_m = -T_{jj}^{-1}u_0, \quad 1 \leqslant j \leqslant M, \tag{2.155}$$

where

$$T_{jm}\sigma_m = \int_{S_m'} g(s, t)\sigma_m(t)dt, \quad s \in S_j.$$

Clearly, $T_{jm} \colon L^2(S_m) \to C^\infty(S_j)$, if $j \neq m$. It is (see [1]) that $T_{jj} \colon L^2(S_j) \to H^1(S_j)$ is an isomorphism if k^2 is not an eigenvalue of the Dirichlet Laplacian in D_j. Since D_j is small, this condition is satisfied. Therefore, $T_{jj}^{-1} \colon H^1(S_j) \to L^2(S_j)$ is a bounded operator and $T_{jj}^{-1}T_{jm}$ is a compact operator from $L^2(S_m)$ into $L^2(S_j)$. Thus, equation (2.155) is of Fredholm type. It is uniquely solvable because the corresponding homogeneous equation has only the trivial solution, as follows from the uniqueness result proved above.

Theorem 2.2.3 is proved. $\square$

Let us derive a formula for the solution. Formula (2.123) implies, as in section 2.2.1, that formula (2.127) holds. The numbers Q_m now can be found by formula similar to equation (2.51):

$$Q_m = -C_m u_e(x_m) := -C_m u_m, \tag{2.156}$$

where C_m is the electrical capacitance of a perfect conductor with the shape D_m, and $u_e(x_m) := u_m$ is the value of the effective field, defined by formula (2.127), and the point $x_m \in D_m$. The numbers u_m can be calculated by solving a linear algebraic system (LAS) similar to equation (2.134):

$$u_j = u_{0j} - \sum_{m \neq j}^{M} g_{jm}C_m u_m, \quad 1 \leqslant m \leqslant M. \tag{2.157}$$

The LAS of the reduced order P, analogous to equation (2.144), is of the form

$$u_q = u_{0q} - \sum_{p \neq q}^{P} g_{pq}C_p a^{-1}N_p u_p |\Delta_p|, \quad 1 \leqslant p \leqslant P. \tag{2.158}$$

Since $C_m = O(a)$, the distribution law, analogous to equation (2.129), takes now the form:

$$\mathcal{N}(\Delta) = \frac{1}{a} \int_\Delta N(x)dx[1 + o(1)], \quad a \to 0. \tag{2.159}$$

Assume for simplicity that all the small bodies are identical. Then

$$C_p a^{-1} = c, \quad c = \text{const} > 0. \tag{2.160}$$

Passing to the limit $a \to 0$ in equation (2.158) yields the integral equation for the limiting effective field

$$u(x) = u_0(x) - c \int_\Omega g(x, y)N(y)u(y)dy. \tag{2.161}$$

This equation is an analog of equation (2.145).

Applying the operator $\nabla^2 + k^2$ to equation (2.161) one gets

$$[\nabla^2 + k^2 - cN(x)]u = 0 \quad \text{in } \mathbb{R}^3. \tag{2.162}$$

This means that

$$n(x) = [1 - k^{-2}cN(x)]^{1/2}. \tag{2.163}$$

This is a formula analogous to equation (2.150).

Let us summarize the results.

Theorem 2.2.4. *Assume equations (2.159), (2.128), and (2.160). Then the solution to problem (2.119), (2.151), (2.121), (2.122) exists, is unique, can be calculated by formula (2.158), where Q_m are calculated by formula (2.156) and u_m solve (2.157).*

As $a \to 0$, the limiting effective field solves equation (2.161). The limiting medium has refraction coefficient (2.163).

2.2.3 The Neumann boundary condition

Let us now consider the scattering problem (2.119)–(2.122) with the boundary condition (2.120) replaced by the Neumann boundary condition

$$u_N|_{S_m} = 0, \quad 1 \leqslant m \leqslant M. \tag{2.164}$$

Theorem 2.2.5. *The above problem has a unique solution. This solution is of the form (2.123). The function $\sigma_m(t)$ in formula (2.123) can be uniquely found from the boundary conditions (2.164).*

Proof. The uniqueness and existence of the solution to problem (2.119), (2.164), (2.121), (2.122) follow from theorem 2.2.1 because $\zeta = 0$ yields this problem. The solution of the form (2.123) can be found: if (2.123) one substitutes into (2.164), one gets a Fredholm-type system of integral equations

$$\frac{A_j\sigma_j - \sigma_j}{2} + \sum_{\substack{m \neq j}}^{M} \frac{\partial T_{jm}\sigma_m}{\partial N_s} = -u_{0N}(s), \quad s \in S_j, \tag{2.165}$$

where

$$T_{jm}\sigma_m := \int_{S_m} \frac{\partial g(s, t)}{\partial N_s}\sigma_m(t)dt, \quad s \in S_j. \tag{2.166}$$

Since the homogeneous equation (2.165) has only the trivial solution, the Fredholm-type system of equations (2.165) is uniquely solvable.

Theorem 2.2.5 is proved. □

The solution u to problem (2.119), (2.164), (2.121), (2.122) can be calculated by formula (2.126) where the numbers Q_m can be calculated by the formula analogous to equation (2.65):

$$Q_m \sim |D_m|\left(\nabla^2 u_e(x_m) + ik\beta_{pq}(x_m)\frac{(x - x_m)_p}{|x - x_m|}\frac{\partial u_e(x_m)}{\partial x_q}\right), \tag{2.167}$$

where $\beta_p := \beta_p(x_m) = \frac{(x - x_m)_p}{|x - x_m|}$, and $(x_m)_p$ is the pth component of the vector x_m, and over the repeated indices p and q one sums up.

Let us assume that the following limits exists:

$$\lim_{a\to 0} \frac{\sum_{x_m\in\Delta_\nu}|D_m|}{|\Delta_\nu|} = \rho(x), \quad x \in \Delta_\nu \tag{2.168}$$

$$\lim_{a\to 0} \frac{\sum_{x_m\in\Delta_\nu}\beta_{pq}(x_m)|D_m|}{|\Delta_\nu|} = B_{pq}(x), \quad x \in \Delta_\nu. \tag{2.169}$$

Here Δ_ν is a partition cube, $|\Delta_\nu|$ is its volume, the side of Δ_ν is equal to b, and assumptions (2.140) and (2.141) hold. The function $\rho(x) > 0$ and $B_{pq}(x)$ are assumed to be smooth.

If these assumptions hold, then the effective field

$$u_e(x) = u_0(x) + \sum_{m\neq j}g(x, x_m)\left(\nabla^2 u_e(x_m) + ik\beta_{pq}(x_m)\frac{(x - x_m)_p}{|x - x_m|}\frac{\partial u_e(x_m)}{\partial x_q}\right)|D_m| \tag{2.170}$$

has a limit $u(x)$ as $a \to 0$, and this limit satisfies the equation:

$$u(x) = u_0(x) + \int_\Omega g(x, y)\left(\rho(y)\nabla^2 u(y) + ikB_{pq}(y)\frac{(x - y)_p}{|x - y|}\frac{\partial u(y)}{\partial y_q}\right)dy. \quad (2.171)$$

This is an integral–differential equation which is not equivalent to a local differential equation in contrast to the scattering problems considered in sections 2.2.1 and 2.2.2.

Let us check that equation (2.171) is of Fredholm type. Apply the operator $\nabla^2 + k^2$ to equation (2.171) and use the equation

$$(\nabla^2 + k^2)g(x, y) = -\delta(x - y) \quad (2.172)$$

to get

$$(\nabla^2 + k^2)u = -\rho(x)\nabla^2 u(x) + ik(\nabla^2 + k^2)$$
$$\int_\Omega g(x, y)B_{pq}(y)\frac{(x - y)_p}{|x - y|}\frac{\partial u(y)}{\partial y_q}dy. \quad (2.173)$$

This equation can be rewritten as

$$\nabla^2 u + \frac{k^2 u}{1 + \rho(x)} = \frac{ik}{1 + \rho(x)}(\nabla^2 + k^2)\int_\Omega g(x, y)\frac{(x - y)_p}{|x - y|}B_{pq}(y)\frac{\partial u(y)}{\partial y_q}dy. \quad (2.174)$$

Let us calculate the expression

$$I := (\nabla_x^2 + k^2)\left(g(x, y)\frac{(x - y)_p}{|x - y|}\right)$$

in the sense of distributions. One has

$$I = -\delta(x - y)\frac{(x - y)_p}{|x - y|} + 2\nabla_x g \cdot \nabla\frac{(x - y)_p}{|x - y|} + g\nabla^2\frac{(x - y)_p}{|x - y|}, \quad (2.175)$$

and

$$\delta(x - y)\frac{(x - y)_p}{|x - y|} = 0, \quad (2.176)$$

$$\nabla_x g \cdot \nabla\frac{(x - y)_p}{|x - y|} = g'\frac{x - y}{|x - y|} \cdot \left(\frac{e_p}{|x - y|} - \frac{(x - y)_p(x - y)}{|x - y|^3}\right) = 0, \quad (2.177)$$

$$g' = \frac{e^{ik|x-y|}}{4\pi|x - y|}\left(ik - \frac{1}{|x - y|}\right). \quad (2.178)$$

Finally,

$$g\nabla^2 \frac{(x-y)_p}{|x-y|} = g\left(-4\pi\delta(x-y)(x-y)_p - 2e_p\frac{x-y}{|x-y|^3}\right) = -2g\frac{x_p-y_p}{|x-y|^3}. \quad (2.179)$$

Thus, in the distributional sense one has:

$$I = -2g\frac{x_p-y_p}{|x-y|^3}. \quad (2.180)$$

Therefore, formula (2.173) can be written as

$$\nabla^2 u + \frac{k^2 u}{1+\rho(x)} = -\frac{2ik}{1+\rho(x)} \int_\Omega \frac{e^{ik|x-y|}(x_p-y_p)}{4\pi|x-y|^4} g(x,y)B_{pq}(y)\frac{\partial u(y)}{\partial y_q}dy. \quad (2.181)$$

The integral in equation (2.181) is a singular integral operator bounded in $L^2(\Omega)$ by theorem X.5.1 in [5, p 257]. Therefore the integral operator in equation (2.181) maps bounded in $L^2(\Omega)$ sets $\frac{\partial u}{\partial y_q}$ into bounded in $L^2(\Omega)$ sets. Since the left-hand side of equation (2.181) is an elliptic operator of the second order and the integral operator in equation (2.181) is bounded as an operator from $H^1(\Omega)$ into $L^2(\Omega)$, it follows that equation (2.181) is of Fredholm type. By $H^1(\Omega) = W^{l,2}(\Omega)$ the Sobolev space is denoted, see [6].

2.2.4 The transmission boundary condition

Consider now the wave scattering problem with the transmission boundary condition:

$$(\nabla^2 + k^2)u = 0 \quad \text{in } D' = \mathbb{R}^3\backslash D, \quad D = \bigcup_{m=1}^{M} D_m, \quad (2.182)$$

$$(\nabla^2 + k_m^2)u = 0 \quad \text{in } D_m, \quad (2.183)$$

$$u^+ = u^-, \quad \rho_m u_N^+ = u_N^- \quad \text{on } S_m, \quad 1 \leqslant m \leqslant M, \quad (2.184)$$

$$u = u_0 + v, \quad \frac{\partial v}{\partial r} - ikv = o\left(\frac{1}{r}\right), \quad r = |x| \to \infty, \quad (2.185)$$

$$(\nabla^2 + k^2)u_0 = 0, \quad u_0 = e^{ik\alpha\cdot x}, \quad \alpha \in S^2. \quad (2.186)$$

We assume $k^2 > 0$, Im $k_m^2 \geqslant 0$, $\rho_m > 0$. As in section 2.1.4 one proves the uniqueness and existence theorem.

Theorem 2.2.6. *Problem (2.182)–(2.186) has a solution.*

Proof. The proof is analogous to the proof of theorems 2.1.10 and 2.1.11. Details are left to the reader, see also [1].

Theorem 2.2.6 is proved. ☐

The solution to the many-body scattering problem (2.182)–(2.186) can be found of the form similar to equation (2.79):

$$u(x) = u_0(x) + \sum_{m=1}^{M} \int_{S_m} g(x, t)\sigma_m(t)dt + \sum_{m=1}^{M} \kappa_m \int_{D_m} g(x, y)u(y)dy, \qquad (2.187)$$

where

$$\kappa_m := k_m^2 - k^2.$$

One verifies, as in section 2.1.4 that the function (2.187) solves equations (2.182), (2.183), (2.185) and the first equation (2.184). Thus, it solves problem (2.182)–(2.186) if it satisfies the second equations (2.184). These equations can be written similarly to equations (2.82) and (2.84):

$$\rho_j \frac{A_j\sigma_j + \sigma_j}{2} - \frac{A_j\sigma_j - \sigma_j}{2} + \sum_{m\neq j}^{M} \int_{S_m} (\rho_j - 1)\frac{\partial g(s, t)}{\partial N}\sigma_m(t)dt$$
$$+ \sum_{m=1}^{M} \kappa_m(\rho_j - 1)\frac{\partial}{\partial N} \int_{D_m} g(x, y)u(y)dy + (\rho_j - 1)u_{0N} = 0. \qquad (2.188)$$

One can use formula (2.117) to write the effective (self-consistent) field acting on the jth small body

$$u_e(x) = u_0(x) + \sum_{m\neq j}^{M} g(x, x_m)\bigg((1 - \rho_m)[\nabla^2 u_e(x_m) - \kappa_m u_e(x_m)]$$
$$+ ik\beta_{pq}(x_m)\frac{\partial u_e}{\partial x_q}\frac{(x - x_m)_p}{|x - x_m|}\bigg)|D_m|$$
$$+ \sum_{m\neq j}^{M} g(x, x_m)\kappa_m u_e(x_m)|D_m|, \qquad |x - x_j| \sim a. \qquad (2.189)$$

Let us derive the integral equation for the limit of the effective field as $a \to 0$ while the assumptions (2.140) and (2.141) hold.

We assume also that

$$\lim_{a\to 0} \frac{\sum_{x_m\in\Delta_p}\rho_m|D_m|}{|\Delta_p|} = \rho(x), \qquad x \in \Delta_p, \qquad (2.190)$$

$$\lim_{a \to 0} \frac{\sum_{x_m \in \Delta_p} \beta_{pq}(x_m)|D_m|}{|\Delta_p|} = B_{pq}(x), \quad x \in \Delta_p, \tag{2.191}$$

$$\lim_{a \to 0} \frac{\sum_{x_m \in \Delta_p} \kappa_m |D_m|}{|\Delta_p|} = K(x), \tag{2.192}$$

and the functions $\rho(x) \geqslant 0$, $B_{pq}(x)$ and $K(x)$ are smooth. Then the limiting equation for $u(x) = \lim_{a \to 0} u_e(x)$ is:

$$u(x) = u_0(x) + \int_\Omega g(x, y) \Bigg[(1 - \rho(y))(\nabla^2 u(y) - K(y)u(y)) $$
$$+ ikB_{pq}(y)\frac{\partial u(y)}{\partial y_q}\frac{x_p - y_p}{|x - y|} + K(y)u(y) \Bigg] dy. \tag{2.193}$$

2.2.5 Wave scattering in an inhomogeneous medium

It is both of theoretical and practical interest to consider wave scattering by one and many small bodies of an arbitrary shape embedded into an inhomogeneous medium. Let us consider this problem for impedance particles. One wants to find the solution to the following problem

$$[\nabla^2 + k^2 n_0^2(x)]u = 0 \quad \text{in } D' = \mathbb{R}^3 \setminus \bigcup_{m=1}^{M} D_m, \tag{2.194}$$

$$\frac{\partial u}{\partial N} = \zeta_m u \quad \text{on } S_m, \quad 1 \leqslant m \leqslant M, \quad \text{Im } \zeta_m \leqslant 0, \tag{2.195}$$

$$u = u_0 + v, \quad \frac{\partial v}{\partial r} - ikv = o\left(\frac{1}{r}\right), \quad r = |x| \to \infty. \tag{2.196}$$

Let us assume that

$$[\nabla^2 + k^2 n_0^2(x)]u_0 = 0 \quad x \in \mathbb{R}^3,$$

the function $n_0^2(x)$ is given and satisfies the conditions:

$$n_0^2(x) = 1, \quad x \in \Omega' = \mathbb{R}^3 \backslash \Omega, \tag{2.197}$$

Ω is a bounded domain where the small bodies are distributed,

$$D := \bigcup_{m=1}^{M} D_m \subset \Omega.$$

Inside Ω, the function $n_0^2(x)$ is an arbitrary continuous function, satisfying the condition

$$\operatorname{Im} n_0^2(x) \geqslant 0, \tag{2.198}$$

which physically means that the medium is not active, that is, it does not generate the energy.

Let $G(x, y)$ be the Green's function satisfying the equation

$$[\nabla^2 + k^2 n_0^2(x)]G(x, y) = -\delta(x - y) \quad \text{in } \mathbb{R}^3, \quad k^2 = \text{const} > 0, \tag{2.199}$$

and the radiation condition

$$\frac{\partial G}{\partial |x|} - ikG = o\left(\frac{1}{|x|}\right), \quad |x| \to \infty. \tag{2.200}$$

Since $n_0^2(x)$ is known, one may assume that $G(x, y)$ is a known function. Of course, if $n_0^2(x)$ is arbitrary then one cannot calculate $G(x, y)$ in a closed form, but one can calculate it numerically.

Let us look for a solution to problem (2.194)–(2.198) of the form

$$u(x) = u_0(x) + \sum_{m=1}^{M} \int_{S_m} G(x, t)\sigma_m(t)dt. \tag{2.201}$$

As in section 2.2.1, one proves that problem (2.194)–(2.198) has a unique solution, and this solution can be found as in equation (2.201). Assume that conditions (2.129), (2.130), (2.140), and (2.141) hold. Then one can repeat all the arguments given in section 2.2.1 and obtain the results similar to theorems 2.2.1, 2.2.2 and lemma 2.2.1.

Note that the asymptotic of the function $G(x, y)$ as $|x - y| \to 0$ is the same as that of $g(x, y)$, and estimates of $G(x, y)$ as $|x - y| \to \infty$ are similar to these of $g(x, y) = \frac{e^{ik|x-y|}}{4\pi |x-y|}$. Let us explain these statements briefly.

One has the following equation for $G(x, y)$:

$$G(x, y) = g(x, y) + \int_\Omega g(x, z)q(z)G(z, y)dz, \tag{2.202}$$

where

$$q(z) := k^2[n_0^2(z) - 1], \quad q(z) = 0, \quad z \in \Omega'. \tag{2.203}$$

Note that

$$g(x, y) = g_0(x, y)[1 + O(|x - y|)], \quad |x - y| \to 0; \quad g_0(x, y) = \frac{1}{4\pi|x - y|}. \tag{2.204}$$

Equation (2.202) is of Fredholm type and its homogeneous version has only the trivial solution (we prove this later), so equation (2.202) has a unique solution. The integral in equation (2.202) is a bounded function.

Therefore,

$$G(x, y) = g_0(x, y)[1 + O(|x - y|)], \quad |x - y| \to 0. \tag{2.205}$$

The solution to equation (2.202) clearly satisfies the radiation condition and the estimate:

$$|G(x, y)| \leq \frac{c}{|x|}, \quad |x| \to \infty, \quad y \in \Omega. \tag{2.206}$$

Let us now prove that the homogeneous version of equation (2.202) has only the trivial solution in $C(\Omega)$. Suppose that

$$w(x) = \int_\Omega g(x, z)q(z)w(z)dz. \tag{2.207}$$

Then w satisfies the radiation condition and solves the equation

$$(\nabla^2 + k^2 + q(x))w = 0, \quad k^2 > 0, \quad \text{Im } q \geq 0, \tag{2.208}$$

where $q(x)$ has compact support.

Multiply equation (2.208) by $\bar{w}$ and the complex conjugate of equation (2.208) by w, subtract from the first equation the second, integrate the result over the ball $B_R := \{x : |x| \leq R\}$, use the Green's formula and the radiation condition for w and get:

$$\lim_{R \to \infty} \left(2ik \int_{|s|=R} |w|^2 ds + 2i \int_\Omega \text{Im } q(x)|w|^2 dx \right) = 0. \tag{2.209}$$

Since $\text{Im } q \geq 0$, it follows from equation (2.209) that

$$\lim_{R \to \infty} \int_{|s|=R} |u|^2 ds = 0. \tag{2.210}$$

Consequently, by the result we have used several times already, $w = 0$ in Ω'. This and the unique continuation theorem for solutions to elliptic equation (2.208) imply that $w = 0$. This completes the proof.

Let us summarize our result.

Theorem 2.2.7. *Problem (2.194)–(2.198) has a unique solution. This solution is of the form (2.201). The effective field in the medium is defined by the formula*

$$u_e(x) = u_0(x) + \sum_{m \neq j} G(x, x_m)Q_m. \tag{2.211}$$

The numbers Q_m are given in formula (2.132) and the numbers u_m in this formula are found by solving linear algebraic system

$$u_j = u_{0j} - \sum_{m \neq j} G_{jm} c_m h_m a^{2-\kappa} u_m, \quad 1 \leqslant j \leqslant M, \tag{2.212}$$

where $G_{jm} := G(x_j, x_m)$, and assumptions (2.128)–(2.130) hold. If $a \to 0$ and assumptions (2.140)–(2.141) hold, then the effective field has a limit $u(x)$ which is the unique solution of the equation

$$u(x) = u_0(x) - c \int_\Omega G(x, y) h(y) N(y) u(y) dy \tag{2.213}$$

where $c > 0$ is a constant defined by the equation $|S| = ca^2$, if one assumes that the small bodies D_m are identical.

The physical interpretation of these results is similar to the one in section 2.2.1. Namely, applying to equation (2.213) the operator $\nabla^2 + k^2 n_j^2(x)$ one gets

$$[\nabla^2 + k^2 n_0^2(x) - cN(x)h(x)]u = 0 \quad \text{in } \mathbb{R}^3. \tag{2.214}$$

Thus

$$n(x) = [n_0^2(x) - ck^{-2}N(x)h(x)]^{1/2}. \tag{2.215}$$

This formula is analogous to equation (2.150).

2.2.6 Summary of the results

Rather than to repeat the formulation of theorems from sections 2.2.1–2.2.5, let us discuss the essential points in a descriptive matter.

In section 2.2 we have given an analytical solution to the many-body wave scattering problems with various boundary conditions and proved that the effective field has a limit, u, as $a \to 0$, and this limit solves equation (2.145) in the case of the impedance boundary condition, and equation (2.161) in the case of the Dirichlet boundary condition. The limiting medium has a new refraction coefficient (2.150) in the case of the impedance boundary condition and a refraction coefficient (2.131) in the case of the Dirichlet boundary condition.

In the case of the Neumann or the transmission boundary conditions the limiting effective field does not satisfy a local differential equation.

The theory developed in this chapter allows one to solve the many-body wave scattering problem for any number of small scatterers. If the number of the scatterers is very large, say, $M > 10^5$, then solving the limiting integral equation, for example, equation (2.145) in the case of the impedance boundary condition, can be a practically efficient way to solve the many-body wave scattering problem.

In chapter 3 we apply the results of section 2.2 to a theory of creating materials with a desired refraction coefficient by embedding in a given material many small particles with prescribed boundary impedances.

References

[1] Ramm A G 1986 *Scattering by Obstacles* (Dordrecht: D. Reidel) pp 1–442
[2] Ramm A G 2013 *Scattering of Acoustic and Electromagnetic Waves by Small Bodies of Arbitrary Shapes. Applications to Creating New Engineered Materials* (New York: Momentum)
[3] Ramm A G 2005 *Wave Scattering by Small Bodies of Arbitrary Shapes* (Singapore: World Scientific)
[4] Bensoussan A, Lions J and Papanicolaou G 2011 *Asymptotic Analysis for Periodic Structures* (Providence, RI: AMS)
[5] Mikhlin S and Prössdorf S 1986 *Singular Integral Operators* (Berlin: Springer)
[6] Gilbarg D and Trudinger N 1983 *Elliptic Partial Differential Equations of Second Order* (New York: Springer)

IOP Publishing

Creating Materials with a Desired Refraction Coefficient
(Second Edition)

Alexander G Ramm

Chapter 3

Creating materials with a desired refraction coefficient

3.1 Scalar wave scattering. Formula for the refraction coefficient

Suppose that in a bounded domain filled with a material with a known refraction coefficient $n_0(x)M$ small $(ka \ll 1)$ bodies D_m, $1 \leqslant m \leqslant M$, of an arbitrary shape are embedded according to the distribution law (2.129), and the boundary impedance of these small bodies is given in (2.130). If conditions (2.140), (2.141), (2.198) hold, then the effective field, acting on the jth body, is given by formula (2.211), where Q_m are defined in formula (2.132) and the numbers $u_m = u_e(x_m)$, $x_m \in D_m$, are calculated by solving linear algebraic system (2.212), in which $G_{jm} = G(x_j, x_m)$ and $G(x, y)$ solves equation (2.199) and satisfies the radiation condition (2.200).

As $a \to 0$, the effective field tends to the unique solution of the integral equation (2.213). The limiting medium has a new refraction coefficient defined by formula (2.215):

$$n(x) = [n_0^2(x) - ck^{-2}h(x)N(x)]^{1/2}. \tag{3.1}$$

The choice of the constant c is in the hands of the experimentalist. This constant is defined by the relation $|S_m| = ca^2$, where S_m is the surface area of a small body D_m and it is assumed that all the small bodies are of the same shape and size. The choice of the functions $N(x) \geqslant 0$ and $h(x)$, $\text{Im } h(x) \leqslant 0$, is also in the hands of the experimentalist.

The square root (3.1) of a complex number $z := n_0^2(x) - ck^{-2}h(x)N(x)$ is defined by the formula:

$$z^{1/2} = |z|^{1/2} e^{i\frac{\phi}{2}}, \quad \phi = \arg z, \quad \phi \in [0, 2\pi). \tag{3.2}$$

If $\text{Im } n_0^2(x) > 0$ and $\text{Im } h(x) \leqslant 0$, then formulas (3.1) and (3.2) give $\arg n(x) \in [0, \frac{\pi}{2})$. If $n_0^2(x) = 0$ and $\text{Im } h = 0$, then formulas (3.1) and (3.2) allow

doi:10.1088/978-0-7503-3391-7ch3

one to get $n(x) \in [0, n_0(x))$. Creating materials with a small refraction coefficient $n(x)$ is of practical interest: it may help to transform curved wave fronts into planar ones and has other possible applications.

By formula (3.1) materials with such a refraction coefficient can be created by embedding into a given material many small impedance particles with prescribed boundary impedances. If the bodies are balls of radius a, then $c = 4\pi$. One can choose $N(x) = $ const, for example, $N(x) = 1$, and then choose

$$h(x) = \frac{n_0^2(x)}{4\pi k^{-2}} \tag{3.3}$$

to get material with $n(x) = 0$.

In the next section we formulate a recipe for creating materials with a desired refraction coefficient.

3.2 A recipe for creating materials with a desired refraction coefficient

Problem 1. *Given a material with known $n_0^2(x)$ in a bounded domain Ω and $n_0^2(x) = 1$ in Ω', one wants to create in Ω a material with a desired refraction coefficient $n^2(x)$.*
Step 1. Given $n_0^2(x)$ and $n^2(x)$, calculate

$$p(x) := k^2[n_0^2(x) - n^2(x)]. \tag{3.4}$$

This is a trivial step.
Step 2. Given $p(x)$ calculate $h(x)$ and $N(x)$ from the equation, see equation (3.1):

$$p(x) = ch(x)N(x). \tag{3.5}$$

The constant c in equation (3.5) one can take to be $c = 4\pi$. In this case the small bodies D_m are balls, and equation (3.5) reduces to

$$p(x) = 4\pi h(x)N(x). \tag{3.6}$$

There are infinitely many solutions h and N to this equation. For example, one can fix arbitrarily $N(x) > 0$ in Ω, $N(x) = 0$ in Ω', and find $h(x) = h_1(x) + ih_2(x)$ by the formulas

$$h_1(x) = \frac{p_1(x)}{4\pi N(x)}, \quad h_2(x) = \frac{p_2(x)}{4\pi N(x)}. \tag{3.7}$$

Here

$$p_1(x) = \operatorname{Re} p(x), \quad p_2(x) = \operatorname{Im} p(x). \tag{3.8}$$

Note that $\operatorname{Im} n^2(x) \geqslant 0$ implies $\operatorname{Im} h(x) \leqslant 0$, so our assumption $\operatorname{Im} h(x) \leqslant 0$ is satisfied. For example, one may take $N(x) = $ const > 0.

Step 2 is also trivial.

Step 3. Given $N(x)$ and $h(x) = h_1(x) + ih_2(x)$, distribute $M = O(\frac{1}{a^{2-\kappa}})$ small impedance balls of radius a in the bounded region Ω according to the distribution law

(2.129), where $\kappa \in [0, 1)$ is the number (parameter) that can be chosen by the experimentalist. The minimal distance d between neighboring particles satisfies condition (2.128). Note that condition $d \gg a$ is satisfied automatically for the distribution law (2.129). Indeed, if d is the minimal distance between neighboring particles, then there are at most $\frac{1}{d^3}$ particles in a cube with the unit side, and since Ω is a bounded domain there are at most $O(\frac{1}{d^3})$ of small particles in Ω. On the other hand, by the distribution law (2.129) one has $\mathcal{N}(\Omega) = O(\frac{1}{a^{2-\kappa}})$. Thus $O(\frac{1}{d^3}) = O(\frac{1}{a^{2-\kappa}})$. Therefore,

$$d = O\left(a^{\frac{2-\kappa}{3}} \right). \tag{3.9}$$

Consequently, condition (2.128) is satisfied. Moreover,

$$\lim_{a \to 0} \frac{a}{d(a)} = 0. \tag{3.10}$$

In the next section we discuss the practical problems one should solve for implementing the above recipe for creating materials with a desired refraction coefficient.

3.3 A discussion of the practical implementation of the recipe

As was mentioned in section 3.2 the first two steps of the recipe are trivial both theoretically and practically. The only step that should be discussed is step 3.

There are two technological problems that must be solved, and if they are solved then practical implementation of our recipe is straightforward.

Technological problem 1: *How does one embed a large number $M = O(\frac{1}{a^{2-\kappa}})$ of small impedance particles into a given material in a bounded domain Ω according to the distribution law (2.129)?*

Technological problem 2: *How does one prepare many small particles with prescribed values of the boundary impedances $\zeta_m = \frac{h(x_m)}{a^\kappa}$?*

Technological problem 1 seems to be solvable currently by the stereolitography process and also by the process of growth of the small particles of desired sizes at the desired points.

Technological problem 2 should be solvable because of the following argument based on the physical reasoning.

If $\zeta = 0$, then the impedance boundary condition becomes the Neumann boundary condition, that is, 'acoustically hard' body satisfies this condition. Thus, such boundary conditions make physical sense and the bodies which satisfy these conditions exist in reality.

The other limiting case of the impedance boundary condition is the Dirichlet boundary condition, corresponding mathematically to the case $\zeta \to \infty$. It corresponds physically to 'acoustically soft' bodies, and such bodies exist in reality. The mathematical limiting process $\zeta \to \infty$ is discussed in [1] in detail.

Thus, the bodies satisfying the impedance boundary condition with any ζ, $\text{Im}\,\zeta \leqslant 0$, should exist.

The general properties of the impedance are restricted physically only by the causality principle, see a discussion of these properties in [2].

It would be of interest in practice to solve technological problem 2.

3.4 Summary of the results

Let us formulate the result in the following theorem.

Theorem 3.4.1. *Given $n_0^2(x)$ and a bounded domain Ω, one can create in Ω a material with a desired refraction coefficient $n(x)$, $\text{Im}\,n^2(x) \geqslant 0$, by embedding $M = O(\frac{1}{a^{2-\kappa}})$ small impedance particles according to the distribution law (2.129). The refraction coefficient $n_a(x)$, corresponding to a finite a, approximates the desired refraction coefficient $n(x)$ in the sense*

$$\lim_{a \to 0} n_a(x) = n(x). \tag{3.11}$$

The functions $h(x)$ and $N(x)$ defining the distribution law (2.129) are found by steps 3.1 and 3.2 of the recipe formulated in section 3.2.

Finally, let us discuss briefly the possibility to create material with negative refraction coefficient. By formulas (3.1) and (3.2) one gets $n(x) < 0$ if the argument of $n_0^2(x) - k^{-2}cN(x)h(x)$ is equal to 2π. Assume that $n_0^2(x) > 0$. We know that $k^{-2}cN(x) \geqslant 0$. Let us take $h(x) = |h(x)|e^{i\varphi}$, where $0 < \varphi$ is very small, that is, $\text{Im}\,h(x) \geqslant 0$ and $\text{Im}\,h(x)$ is very small. We have proved in section 2.2.1 that equation (2.145) is uniquely solvable for $\text{Im}\,h \geqslant 0$ sufficiently small. For such $h(x)$ one has $\arg(n_0^2(x) - k^{-2}cN(x)h(x))$ is very close to 2π and the square root (3.1) is negative, provided that $\text{Re}(n_0^2(x) - k^{-2}cN(x)h(x)) > 0$. This argument shows that it is possible to create materials with negative refraction coefficient $n(x)$ by embedding in a given material many small particles with properly chosen boundary impedances.

References

[1] Ramm A G 2013 *Scattering of Acoustic and Electromagnetic Waves by Small Bodies of Arbitrary Shapes. Applications to Creating New Engineered Materials* (New York: Momentum)
[2] Landau L and Lifshitz E 1984 *Electrodynamics of Continuous Media* (Oxford: Pergamon)

IOP Publishing

Creating Materials with a Desired Refraction Coefficient (Second Edition)

Alexander G Ramm

Chapter 4

Wave-focusing materials

4.1 What is a wave-focusing material?

Consider the scattering problem

$$[\nabla^2 + k^2 - q(x)]u = 0 \quad \text{in } \mathbb{R}^3. \tag{4.1}$$

The scattering solution to equation (4.1) is of the form

$$u(x, \alpha, k) = e^{ik\alpha \cdot x} + A(\beta, \alpha, k)\frac{e^{ikr}}{r} + o\left(\frac{1}{r}\right),$$

$$r = |x| \to \infty, \quad \frac{x}{r} := \beta. \tag{4.2}$$

Here $\alpha \in S^2$ is a given unit vector in the direction of propagation of the incident plane wave, the coefficient $A(\beta, \alpha, k) = A_q(\beta, \alpha, k)$ is called the scattering amplitude, β is the unit vector in the direction of the scattered wave. The function $A(\beta, \alpha, k)$ describes the angular distribution of the scattered field. The cross-section

$$\sigma(\alpha) = \int_{S^2} |A(\beta, \alpha, k)|^2 d\beta \tag{4.3}$$

describes the total energy scattered by the potential q when the incident field $e^{ik\alpha \cdot x}$ is incident upon the potential. If $q(x)$ is compactly supported, $q \in L^2$, and Im $q \leqslant 0$, then the scattering solution to equation (4.1) exists and is unique. This is proved as in the proof of theorem 2.2.7.

From the mid-forties of the twentieth century physicists wanted to prove that the observable quantities determine a physical system, that is, determine its Hamiltonian. For non-relativistic quantum-mechanical scattering the potential

doi:10.1088/978-0-7503-3391-7ch4

describes the Hamiltonian, and the scattering amplitude $A(\beta, \alpha, k)$ is 'the observable'. This brings the basic question:

Given the scattering amplitude, can one uniquely determine the potential q in equation (4.1)?

This question is not yet quite clearly stated because it is not specified for what β, α and k the scattering amplitude is known.

If $A(\beta, \alpha, k)$ is known for all β and α on the unit sphere and for all $k > 0$, then it is easy to prove that q is uniquely determined by these data (see, for example, [1]).

If the scattering amplitude is known at a fixed $k > 0$ (a fixed energy $k^2 > 0$) and q is a compactly supported real-valued potential, $q \in L^2(\mathbb{R}^3)$, then the uniqueness of finding such a q from the fixed-energy scattering data was announced in [2] and proved in [3], see also [1, 4–8].

An algorithm for reconstruction of q from fixed-energy scattering data was presented by the author in [1, 9]. For many decades the uniqueness of the solution to inverse scattering problem of finding the potential from non-overdetermined scattering data was open. By non-overdetermined scattering data one understands the value of the scattering set of the parameters β, α, k, for example, back-scattering data $A(-\alpha, \alpha, k)$, $\alpha \in S^2, \forall k > 0$, or the fixed incident direction data $A(\beta, \alpha_0, k)$, $\forall \beta \in S^2, \forall k > 0$, and a fixed $\alpha = \alpha_0 \in S^2$. Note that the unknown potential q depends on these variables.

The uniqueness of the solution to the inverse scattering problem with non-overdetermined data was first announced in [10] and proved in [11–13].

The inverse scattering problem with underdetermined data was not studied in the literature. By underdetermined scattering data we understand the values of $A(\beta, \alpha, k)$ on a set of dimension less than three. For example, consider the scattering data

$$A_q(\beta) := A(\beta) := A(\beta, \alpha_0, k_0), \quad \forall \beta \in S^2, \tag{4.4}$$

that is, the values of the scattering amplitude at a fixed $k = k_0 > 0$ and, a fixed $\alpha = \alpha_0 \in S^2$, and all $\beta \in S^2$.

Inverse scattering problem (IP): *Given an arbitrary function $f(\beta) \in L^2(S^2)$ and an arbitrary small number $\varepsilon > 0$, can one find a potential $q \in L^2(\Omega)$ such that*

$$\|A_q(\beta) - f(\beta)\| \leqslant \varepsilon, \tag{4.5}$$

where the norm $\|\cdot\|$ is $L^2(S^2)$ norm?

We give a positive answer to this question and construct a potential $q \in L^2(\Omega)$ that generates the scattering data $A_q(\beta)$ satisfying (4.5).

A priori it is not clear that such a potential does exist. By Ω we mean an arbitrary but fixed bounded domain outside of which $q = 0$.

Note that the inverse scattering problem with underdetermined scattering data in general has many solutions. Therefore, we look for a solution. Since such problems were not studied earlier, there are open questions in the theory. Some of these questions will be formulated in this chapter.

In the engineering literature the scattering amplitude $A(\beta, \alpha, k)$ is often called the radiation pattern.

A material with a wave-focusing property is a material which generates the scattering amplitude $A_q(\beta)$ satisfying inequality (4.5) in which $f(\beta)$ vanishes outside of a prescribed small solid angle.

The relation between the potential $q(x)$ and the refraction coefficient of the material is given by the relation

$$k^2 n^2(x) = k^2 - q(x), \quad n^2(x) = 1 - k^{-2}q(x). \tag{4.6}$$

In other words,

$$q(x) = k^2(1 - n^2(x)). \tag{4.7}$$

It will be convenient in this chapter to deal with potentials q rather than with the refraction coefficients n, although there is a simple relation between these functions.

4.2 Creating wave-focusing materials

Throughout we assume that Ω is a bounded domain in $\mathbb{R}^3$.

Our first result is the following theorem.

Theorem 4.2.1. *Let $f(\beta) \in L^2(S^2)$ be an arbitrary fixed given function, and $\varepsilon > 0$ be an arbitrary small number. Then there exists a $q \in L^2(\Omega)$ such that inequality (4.5) holds.*

Proof. The proof will require several lemmas.

Lemma 4.2.1. *Suppose that the set $\{A_q(\beta)\}$ is dense in $L^2(S^2)$ when q runs through $L^2(\Omega)$. Then the set $\{A_q(\beta)\}$ is dense in $L^2(S^2)$ when q runs through $C_0^\infty(\Omega)$.*

Proof of lemma 4.2.1. The proof is based on the following formula, proved by the author [1, p 262], [9]. This result is of independent interest:

$$-4\pi[A_{q_1}(\beta, \alpha, k) - A_{q_2}(\beta, \alpha, k)] = \int_\Omega [q_1(x) - q_2(x)]u_1(x, \alpha)u_2(x, -\beta)dx, \tag{4.8}$$

where $u_j(x, \alpha)$ is the scattering solution corresponding to the potential q_j, $j = 1, 2$.

We give a proof of formula (4.8) below.

Assuming that formula (4.8) is proved and also that the following inequality

$$\sup_{\alpha \in S^2} \sup_{x \in \mathbb{R}^3} |u_j(x, \alpha)| \leqslant c, \tag{4.9}$$

is established, let us prove lemma 4.2.1.

A proof of inequality (4.9) is given below.

To prove lemma 4.2.1 note that equations (4.8) and (4.9) imply

$$\|A_{q_1}(\beta) - A_{q_2}(\beta)\|_{L^2(S^2)} \leqslant c\|q_1 - q_2\|_{L^2(\Omega)}. \tag{4.10}$$

Therefore, small variations of q in $L^2(\Omega)$ norm lead to small variations of $A_q(\beta)$ in $L^2(S^2)$ norm.

Since the set $C_0^\infty(\Omega)$ is dense in $L^2(\Omega)$, it follows that if a function $f \in L^2(S^2)$ can be approximated with an arbitrary accuracy in $L^2(S^2)$ by a function $A_q(\beta)$ with some $q_1 \in L^2(\Omega)$, then there is a potential $q_2 \in C_0^\infty(\Omega)$ which approximates q_1 with an arbitrary accuracy in the norm of $L^2(\Omega)$. By inequality (4.10) the function $A_{q_2}(\beta)$ approximates with any desired accuracy the original function $f(\beta)$ in $L^2(S^2)$.

Lemma 4.2.1 is proved. $\qquad\qquad\qquad\qquad\qquad\qquad\qquad\qquad\qquad\qquad\quad\square$

Lemma 4.2.2. *Formula (4.8) holds.*

Proof. In this proof we assume $k > 0$ arbitrary and fixed, and denote by $G(x, y)$ the resolvent kernel of the operator $-\nabla + q - k^2$, that is

$$(-\nabla + q(x) - k^2)G(x, y) = \delta(x - y), \quad \text{in } \mathbb{R}^3, \tag{4.11}$$

where G satisfies the radiation condition.

The uniquely solvable integral equation for G is

$$G(x, y) = g(x, y) - \int_\Omega g(x, z)q(z)G(z, y)dz, \quad g(x, y) = \frac{e^{ik|x-y|}}{4\pi|x - y|}. \tag{4.12}$$

Note that

$$g(x, y) = \frac{e^{ik|x|}}{4\pi|x|}e^{-ik\beta\cdot y}\left[1 + O\left(\frac{1}{|x|}\right)\right], \quad |x| \to \infty, \quad \beta := \frac{x}{|x|}, \tag{4.13}$$

where y is any vector in a bounded domain and $e^{-ik\beta\cdot y}$ is the scattering solution $u_0(y, -\beta, k)$ for $q = 0$. One can prove, using equation (4.12), that

$$G(x, y) = \frac{e^{ik|x|}}{4\pi|x|}u(y, -\beta, k)\left(1 + O\left(\frac{1}{|x|}\right)\right), \quad |x| \to \infty, \quad \beta := \frac{x}{|x|}, \tag{4.14}$$

where $u(y, -\beta, k)$ is the scattering solution, that is

$$(\nabla^2 + k^2 - q(x))u(x, -\beta, k) = 0 \quad \text{in } \mathbb{R}^3, \tag{4.15}$$

$$u(x, -\beta, k) = e^{ik\beta\cdot x} + v, \tag{4.16}$$

where v satisfies the radiation condition.

The existence of the representation (4.14) follows from equation (4.12) and formula (4.13). One can derive the following formula

$$u(y, -\beta, k) = e^{ik\beta\cdot y} - \int_\Omega e^{-ik\beta\cdot z}q(z)G(z, y)dz, \tag{4.17}$$

from equations (4.12) and (4.13). The fact that function (4.17) is the scattering solution can be checked easily: the integral in equation (4.17) satisfies the radiation condition because $G(z, y)$ satisfies this condition, and the function (4.17) satisfies equation (4.15) because

$$
\begin{aligned}
(\nabla^2 + k^2)u(y, -\beta, k) &= -\int_\Omega e^{-ik\beta \cdot z} q(z)[q(y)G(z, y) - \delta(z - y)]dz \\
&= q(y)[u(y, -\beta, k) - e^{-ik\beta \cdot y}] + e^{-ik\beta \cdot y}q(y) \\
&= q(y)u(y, -\beta, k).
\end{aligned}
\tag{4.18}
$$

Since problem (4.15) and (4.16) has unique solution, it follows that function (4.17) is the scattering solution.

To prove formula (4.8) we start with the formula

$$
G_{q_2}(x, y) = G_{q_1}(x, y) - \int_\Omega G_{q_1}(x, z)[q_2(z) - q_1(z)]G_{q_2}(z, y)dz,
\tag{4.19}
$$

that is verified easily by applying the operator $\nabla_x^2 + k^2 - q_1(x)$ to both sides of the formula (4.19). Let us take $|x| \to \infty$, $\frac{x}{|x|} = \beta$, and use formula (4.14) to get

$$
u_2(y, -\beta, k) = u_1(y, -\beta, k) - \int_\Omega u_1(z, -\beta, k)[q_2(z) - q_1(z)]G_{q_2}(z, y)dz.
\tag{4.20}
$$

Now, let us take $|y| \to \infty$, $\frac{y}{|y|} = -\alpha$, and use formula (4.14) and formula (4.2) to get:

$$
\begin{aligned}
A_2(-\alpha, -\beta, k) = A_1(-\alpha, -\beta, k) \\
-\frac{1}{4\pi} \int [q_2(z) - q_1(z)]u_1(z, -\beta, k)u_2(z, \alpha, k)dz.
\end{aligned}
\tag{4.21}
$$

Finally, take into account the known relation the reciprocity property (see [14]):

$$
A(-\alpha, -\beta, k) = A(\beta, \alpha, k),
\tag{4.22}
$$

and get from equation (4.21) the desired formula (4.8).

Lemma 4.2.2 is proved. $\qquad\square$

Lemma 4.2.3. *Formula (4.9) holds.*

Proof of lemma 4.2.3. The scattering solution solves the equation

$$
u(x, \alpha, k) = e^{ik\beta \cdot x} - \int_\Omega g(x, y)q(y)u(y, \alpha, k)dy := u_0 - Tu,
\tag{4.23}
$$

where the operator $T: C(\Omega) \to C(\Omega)$ and $(I + T)^{-1}$ is bounded. Therefore,

$$
\sup_{x \in \Omega}|u(x, \alpha, k)| \leqslant c \sup_{x \in \Omega}|u_0(x, \alpha, k)|,
\tag{4.24}
$$

where the constant c does not depend on $\alpha \in S^2$.

Consequently, estimate (4.9) follows.

Lemma 4.2.3 is proved. $\qquad\square$

This completes the proof of lemma 4.2.1.

Remark 4.2.1. *The results of lemmas 4.2.1–4.2.3 belong to the author (see [1] and [14]).*

Let us continue with the proof of theorem 4.2.1.

Lemma 4.2.4. *The set $\{A_q(\beta)\}$ is dense in $L^2(S^2)$ when q runs through $C_0^\infty(\Omega)$.*

Proof of lemma 4.2.4. Suppose the conclusion of lemma 4.2.4 is false and derive a contradiction from this assumption.

If the conclusion of lemma 4.2.4 is false then there exists a function $f(\beta) \in L^2(S^2)$, $f \not\equiv 0$, such that

$$\int_{S^2} f(\beta)A_q(\beta)d\beta = 0, \quad \forall q \in C_0^\infty(\Omega). \tag{4.25}$$

Recall that

$$A_q(\beta) = -\frac{1}{4\pi} \int_\Omega e^{-ik\beta \cdot y}q(y)u(y, \alpha, k)dy. \tag{4.26}$$

From equation (4.25) it follows that

$$\int_\Omega dy q(y)u(y, \alpha, k) \int_{S^2} f(\beta)e^{-ik\beta \cdot y}d\beta = 0, \quad \forall q \in C_0^\infty(\Omega). \tag{4.27}$$

It follows from equation (4.27) and from the sufficient arbitrariness of q that

$$\int_{S^2} f(\beta)e^{-ik\beta \cdot y}d\beta = 0, \quad \forall y \in \Omega. \tag{4.28}$$

Here we have used the fact that the scattering solutions do not vanish on an open subset of $\mathbb{R}^3$ as follows from the unique continuation property of the solutions to homogeneous Schrödinger equations.

It follows from equation (4.28) that $f = 0$. Indeed, the left-hand side of equation (4.28) is Fourier transform of a compactly supported distribution supported on the sphere of radius $k > 0$ with the density $f(\beta)$. This Fourier transform is an entire function of y which vanishes on an open subset $\Omega \subset \mathbb{R}^3$. Therefore, it vanishes identically in $\mathbb{R}^3$. By the injectivity of the Fourier transform one concludes that $f(\beta) = 0$.

Lemma 4.2.4 is proved. □

From lemma 4.2.4 the conclusion of theorem 4.2.1 follows. □

The question of interest now is to find a $q \in L^2(\Omega)$ such that inequality (4.5) holds. The existence of such a q follows from theorem 4.2.1. Finding such a q amounts to solving the underdetermined inverse scattering problem that we have discussed in section 4.1.

If $f(\beta) \in L^2(S^2)$ is an arbitrary given function, then one finds some $h(y) \in L^2(D)$ such that

$$\left\| f(\beta) + \frac{1}{4\pi} \int_\Omega h(y) e^{-ik\beta \cdot y} dy \right\|_{L^2(S^2)} \leqslant \varepsilon. \qquad (4.29)$$

Here $h(y) := q(y)u(y, \alpha, k)$, see formula (4.26). There are many $h \in L^2(\Omega)$ for which (4.29) holds. Indeed, if inequality (4.29) holds for some h, then it holds with 2ε in place of ε for any h_δ such that $\|h_\delta - h\|_{L^2(\Omega)} \leqslant \delta$, if $\delta > 0$ is sufficiently small.

We will discuss methods for finding $h \in L^2(\Omega)$ satisfying inequality (4.29) in section 4.3.

If h found from equation (4.29) then one would like to find a $q \in L^2(\Omega)$ such that

$$\|h(x) - qu(x, \alpha, k)\|_{L^2(\Omega)} \leqslant \delta, \qquad (4.30)$$

where $\delta > 0$ is an arbitrary small number, and $\alpha \in S^2$ is fixed. Recall, that $k > 0$ is fixed also.

It is possible to satisfy inequality (4.30). This is proved in the following theorem.

Theorem 4.2.2. *The set* $\{q(x)u_q(x, \alpha, k)\}_{\forall q \in L^2(\Omega)}$ *is dense in* $L^2(\Omega)$ *and* $u_q(x, \alpha, k) := u(x, \alpha, k)$ *is the scattering solution corresponding to* q, *where* q *is a real-valued compactly supported potential.*

Proof. We assume that $\alpha \in S^2$ is fixed, $k > 0$ is fixed, and denote

$$q(x)u_q(x, \alpha, k) := h(x). \qquad (4.31)$$

We want to prove that the set $\{h(x)\}_{\forall q \in L^2(\Omega)}$ is dense in $L^2(\Omega)$. Note that if $u_0(x) = e^{ik\alpha \cdot x}$ and h is of the form (4.31), then

$$q(x) = \frac{h(x)}{u_q(x, \alpha, k)} = \frac{h(x)}{u_0(x) - \int_\Omega g(x, y)h(y)dy}. \qquad (4.32)$$

If $h \in L^2(\Omega)$ is arbitrary, then formula (4.32) defines some function $q(x)$ which may not belong to $L^2(\Omega)$. Since the denominator in (4.32) is in $H^2_{loc}(\mathbb{R}^3)$ if $h \in L^2(\Omega)$, the function $q(x)$ defined by formula (4.32) fails to be in $L^2(\Omega)$ if and only if the denominator vanishes at some set in $\mathbb{R}^3$. By $H^2_{loc}(\mathbb{R}^3)$ the Sobolev space of functions that belong together with their derivatives up to the order two to L^2 on every compact subset of $\mathbb{R}^3$ is denoted.

If the denominator $u = u_q$ in (4.32) vanishes at some set of points in Ω, then

$$\text{Re } u(x) = 0, \quad \text{Im } u(x) = 0. \qquad (4.33)$$

These are two equations for twice differentiable functions Re u and Im u of $x = (x_1, x_2, x_3)$. These two equations define a line l in $\mathbb{R}^3$. Thus, the set of zeros of u is the set of measure zero in $\mathbb{R}^3$. Generally, the line l is smooth. Indeed, since we want to prove that the set of h of the form (4.32) is dense in $L^2(\Omega)$, one may assume $h \in C_0^\infty(\Omega)$, and in this case the functions Re $u(x)$ and Im $u(x)$ are $C^\infty(\Omega)$ functions and l is smooth. The vectors $\nabla \text{Re}\, u$ and $\nabla \text{Im}\, u$ can be considered linearly independent and in this case the line l, defined by the equations (4.33) is smooth by the implicit function theorem. Recall that $\nabla \text{Re}\, u$ and $\nabla \text{Im}\, u$ are vectors directed along normals to the surfaces (4.33). We want to cover the line l, that is, the set of zeros of u, by a tubular neighborhood

$$N_\delta := \{x \colon x \in \Omega, |u(x)| \leqslant \delta\}, \quad \Omega_\delta := \Omega \backslash N_\delta. \tag{4.34}$$

Choose

$$h_\delta = \begin{cases} h & \text{in } \Omega_\delta, \\ 0 & \text{in } N_\delta. \end{cases} \tag{4.35}$$

Then

$$q_\delta = \frac{h_\delta(x)}{u_0(x) - \int_\Omega g(x, y) h_\delta(y) dy} = \begin{cases} \dfrac{h_\delta(x)}{u_0(x) - \int_{\Omega_\delta} g(x, y) h_\delta(y) dy} & \text{in } \Omega_\delta, \\ 0 & \text{in } N_\delta. \end{cases} \tag{4.36}$$

Since the measure of N_δ is arbitrarily small if δ is sufficiently small, one has

$$\|h - h_\delta\|_{L^2(\Omega)} = o(1), \quad \delta \to 0. \tag{4.37}$$

Therefore, if equation (4.29) holds for h, then it will hold for h_δ with 2ε replacing ε, provided that $\delta > 0$ is sufficiently small.

Let us check that q_δ, defined in equation (4.36), belongs to $L^\infty(\Omega)$. One has

$$
\begin{aligned}
|q_\delta(x)| &\leqslant \frac{\max_{x\in\Omega} |h_\delta(x)|}{\min_{x\in\Omega_\delta} \left| u_0(x) - \int_\Omega g(x, y) h_\delta(y) dy \right|} \\
&\leqslant c \left(\min_{x\in\Omega_\delta} \left| u_0(x) - \int_\Omega g(x, y) h(y) dy \right| \right. \\
&\quad \left. - \max_{x\in\Omega_\delta} \left| u_0(x) - \int_\Omega g(x, y)|h(y) - h_\delta(y)| dy \right| \right)^{-1}.
\end{aligned}
\tag{4.38}
$$

By definition (4.34) one has

$$\min_{x\in\Omega_\delta} \left| u_0(x) - \int_\Omega g(x, y) h(y) dy \right| = \min_{x\in\Omega_\delta} |u(x)| > \delta. \tag{4.39}$$

One has

$$\int_\Omega g(x, y)|h(y) - h_\delta(y)| dy \leqslant \int_{N_\delta} \frac{|h(y) - h_\delta(y)|}{4\pi|x - y|} dy \leqslant c \int_{N_\delta} \frac{dy}{|x - y|}, \tag{4.40}$$

where

$$c = \frac{\max_{y\in\Omega} |h(y)|}{4\pi}. \tag{4.41}$$

Let us prove that

$$I_\delta := \int_{N_\delta} \frac{dy}{|x - y|} \leqslant c\delta^2|\ln \delta|, \quad x \in \Omega_\delta, \quad \delta \to 0, \tag{4.42}$$

where $c > 0$ is a constant. By c we denote various constants independent of δ.

If equation (4.42) is proved, then

$$|q_\delta(x)| \leqslant \frac{c}{\delta - c\delta^2|\ln \delta|} = O\left(\frac{1}{\delta}\right), \tag{4.43}$$

if δ is sufficiently small.

To prove equation (4.42) let us use the coordinates s_1, s_2, s_3, such that s_1 is directed along the normal to the surface $u_1 := \mathrm{Re}\, u = 0$, s_2 is directed along the normal to the surface $u_2 := \mathrm{Im}\, u = 0$, s_3 is directed along the tangent vector to the curve l, the origin of the coordinate system is located in l, and the s_1, s_2 axes are orthogonal to s_3 axis, that is s_3 axis is orthogonal to the plane in which axes s_1 and s_2 lie.

The Jacobian J of the transformation $(x_1, x_2, x_3) \to (s_1, s_2, s_3)$ at the origin is

$$J = \frac{\partial(u_1, u_2, x_3)}{(x_1, x_2, x_3)}\bigg|_{x=0} = \begin{vmatrix} u_{1,1} & u_{1,2} & 0 \\ u_{2,1} & u_{2,2} & 0 \\ 0 & 0 & 1 \end{vmatrix} \neq 0, \tag{4.44}$$

where $u_{i,j} := \frac{\partial u_i}{\partial x_j}$, and we took into account that

$$\begin{vmatrix} u_{1,1} & u_{1,2} \\ u_{2,1} & u_{2,2} \end{vmatrix} \neq 0,$$

because the normals to the surfaces $u_1 = 0$ and $u_2 = 0$ are linearly independent, that is, ∇u_1 and ∇u_2 are linearly independent and are orthogonal to the line $x_3 = s_3$ at the origin.

In a neighborhood of the origin one has in N_δ the inequality

$$|J| + |J^{-1}| \leqslant c. \tag{4.45}$$

The integral (4.42) can be written in the coordinates s_1, s_2, s_3, and estimated as follows:

$$I_\delta \leqslant c \int_0^\delta \rho\, d\rho \int_0^{c_3} \frac{ds_3}{\sqrt{\rho^2 + s_3^2}}, \tag{4.46}$$

because the domain of integration can be described by the inequalities

$$u_1^2 + u_2^2 \leqslant \delta^2, \quad 0 \leqslant s_3 \leqslant c_3, \tag{4.47}$$

where $c_3 > 0$ is some constant. We have also used the inequalities

$$c_1(u_1^2 + u_2^2 + s_3^2) \leqslant |y|^2 \leqslant c_2(u_1^2 + u_2^2 + s_3^2), \tag{4.48}$$

which hold because of equation (4.45).

One has

$$\int_0^{c_3} \frac{ds_3}{\sqrt{\rho^2 + s_3^2}} = \ln\left(s_3 + \sqrt{\rho^2 + s_3^2}\right)\Big|_0^{c_3} \leqslant c_4 + \ln\frac{1}{\rho} \leqslant c_5 \ln\frac{1}{\rho}. \qquad (4.49)$$

Therefore estimate (4.42) follows from equations (4.46) and (4.49). Consequently, inequality (4.38) yields

$$|q_\delta(x)| \leqslant \frac{c}{\delta - O(\delta^2|\ln\delta|)} = O\left(\frac{1}{\delta}\right) \qquad (4.50)$$

Therefore, if formula (4.32) does not give $q \in L^2(\Omega)$ then a small perturbation of h, defined in equation (4.35) yields a function q_δ bounded in Ω.

We have proved that the set $\{h_\delta\} = \{q_\delta u_\delta\}$ is dense in $L^2(D)$, where q_δ are bounded functions.

This proves theorem 4.2.2. $\qquad\qquad\qquad\qquad\qquad\qquad\qquad\qquad\qquad\qquad$ □

Remark 4.2.2. *Let $H_0^2(\Omega)$ be the Sobolev space of twice differentiable functions whose second derivatives belong to $L^2(\Omega)$ and the functions vanish in a neighborhood of the boundary $\partial\Omega$ of Ω. Then*

$$\int_\Omega e^{-ik\beta\cdot x}(\nabla^2 + k^2)\Phi dx = 0 \quad \forall\Phi \in H_0^2(\Omega), \qquad (4.51)$$

as one can check by using Green's formula and taking into account that

$$\Phi = \frac{\partial\Phi}{\partial N} = 0 \quad on\ \partial\Omega. \qquad (4.52)$$

Therefore,

$$\int_\Omega g(x, y)[h(y) + (\nabla^2 + k^2)\Phi]dy = \int_\Omega g(x, y)h(y)dy - \Phi(x), \qquad (4.53)$$

where we have used Green's formula, taken into account the equation $(\nabla^2 + k^2)g = -\delta(x - y)$, and the boundary conditions (4.52).

Consequently, the potentials $q(x)$ and $q_\Phi(x)$ generate the same scattering amplitude.

Here $\Phi \in H_0^2(\Omega)$ and

$$q_\Phi(x) = \frac{h(y) + (\nabla^2 + k^2)\Phi}{u_0 - \int_\Omega g(x, y)h(y)dy + \Phi(x)}, \quad q(x) = \frac{h(y)}{u_0 - \int_\Omega g(x, y)h(y)dy} \qquad (4.54)$$

One may try to use this property in order to find Φ such that $\text{Im}\,q_\Phi \leqslant 0$. This is of interest because the inequality $\text{Im}\,q_\Phi \leqslant 0$ guarantees the uniqueness of the scattering solutions. The problem of the existence of such Φ is currently open.

4.3 Computational aspects of the problem

There are two problems that we discuss from the computational point of view.
The first problem consists of finding $h(x) \in \Omega$ such that

$$\left\| f(\beta) + \frac{1}{4\pi} \int_{\Omega} e^{-ik\beta \cdot y} h(y) dy \right\|_{L^2(S^2)} \leq \varepsilon. \tag{4.55}$$

Here $f \in L^2(S^2)$ and a small number $\varepsilon > 0$ are given, and $h \in L^2(\Omega)$ is to be found.
Existence of such h is proved in theorem 4.2.1. There are many h for which
inequality (4.55) holds. Let us give an algorithm for finding such an h. Let us assume
without loss of generality that Ω is a ball $B_R := B$ centered at the origin and of radius
R. This can be assumed without loss of generality because one can always choose
$B \subset \Omega$ and if $h \in L^2(B)$ is found which satisfies inequality (4.55), then this h can be
extended to $\Omega \backslash B$ by zero, it will be then defined in Ω, it will belong to $L^2(\Omega)$, and it
will satisfy inequality (4.55).

Expand the plane wave $e^{-ik\beta \cdot y}$ and $h(y)$ into the spherical harmonic series (see, for
example, [1]):

$$e^{-ik\beta \cdot y} = \sum_{l=0}^{\infty} 4\pi i^l j_l(kr) Y_l(-\beta) \overline{Y_l(y^0)}, \quad r := |y|, \quad y^0 := \frac{y}{r}, \tag{4.56}$$

$$j_l(r) := \left(\frac{\pi}{2r}\right)^{1/2} J_{l+\frac{1}{2}}(r), \quad \frac{4\pi}{2l+1} \sum_{m=-l}^{l} Y_{l,m}(x^0) \overline{Y_{l,m}(y^0)} = P_l(x^0 \cdot y^0). \tag{4.57}$$

Here $J_{l+\frac{1}{2}}(r)$ is the Bessel function regular at $r = 0$, $Y_l(\beta) = Y_{lm}(\beta)$, $-l \leq m \leq l$, are
the spherical harmonics, the over bar $\overline{Y_l(y^0)}$ stands for complex conjugate,

$$Y_{l,m}(-\beta) = (-1)^l Y_{l,m}(\beta), \quad \overline{Y_{l,m}(\beta)} = (-1)^{l+m} Y_{l,-m}(\beta), \tag{4.58}$$

$$Y_{l,m}(\alpha) := \frac{(-1)^{m+|m|} i^l}{\sqrt{4\pi}} \left[\frac{(2l+1)(l-|m|)!}{(l+|m|)!} \right]^{1/2} e^{im\phi} P_{l,m}(\cos\theta), \tag{4.59}$$

$$P_{l,m}(\cos\theta) := (\sin\theta)^m \frac{d^m P_l(\cos\theta)}{(d\cos\theta)^m}, \quad P_l(t) := \frac{1}{2^l l!} \frac{d^l(t^2-1)^l}{dt^l}, \tag{4.60}$$

where $t = \cos\theta$, $P_l(t)$ are Legendre polynomials, $P_{l,m}(\cos\theta)$ are associated Legendre
functions (see [15]), the unit vector β is described by the spherical coordinates (θ, ϕ),
$0 \leq \theta \leq \pi$, $0 \leq \phi \leq 2\pi$, the summation in (4.56) and below $\sum_{l=0}^{\infty}$ is understood as
$\sum_{l=0}^{\infty} \sum_{m=-l}^{l}$, and

$$(Y_{l,m}, Y_{l',m'})_{L^2(S^2)} = \delta_{ll'} \delta_{mm'}. \tag{4.61}$$

Let

$$h(y) = \sum_{l=0}^{\infty} c_l(r) Y_l(y^0), \quad r = |y|, \quad y^0 = \frac{y}{r}, \tag{4.62}$$

where the Fourier coefficients $c_l(r)$ are defined by the formula:

$$c_l(r) := \int_{S^2} h(y) \overline{Y_l(y^0)} dy^0. \tag{4.63}$$

Let

$$f_L(\beta) = \sum_{l=0}^{L} f_l Y_l(\beta), \tag{4.64}$$

where L is a sufficiently large integer such that

$$\|f - f_L\|_{L^2(S^2)} < \frac{\varepsilon}{2}. \tag{4.65}$$

Thus, if

$$\left\| f_L(\beta) + \frac{1}{4\pi} \int_B e^{-ik\beta \cdot y} h(y) dy \right\| < \frac{\varepsilon}{2}, \tag{4.66}$$

then inequality (4.55) holds.

Let us find an h satisfying inequality (4.66). Consider the equation for h_l:

$$\int_B e^{-ik\beta \cdot y} h_L(y) dy = -4\pi f_L(\beta). \tag{4.67}$$

Substitute equations (4.62) and (4.64) into equation (4.67) and use formulas (4.56), (4.58) and (4.61) to get

$$(-i)^{l+2} \int_0^R r^2 j_l(kr) c_l(r) dr = f_l, \tag{4.68}$$

where $c_l(r) = c_{l,m}(r)$, $f_l = f_{l,m}$. Equation (4.68) has many solutions. Let $c_l^{\perp}(r)$ be any function orthogonal to $r^2 j_l(kr)$, that is,

$$\int_0^R r^2 j_l(kr) c_l^{\perp}(r) dr = 0. \tag{4.69}$$

The general solution to equation (4.68) is

$$c_l(r) = a_l f_l j_l(kr) + b_l c_l^{\perp}(r), \tag{4.70}$$

where b_l is an arbitrary constant and

$$a_l = \frac{1}{(-i)^{l+2} \int_0^R r^2 j_l^2(kr) dr}. \tag{4.71}$$

We have proved the following theorem:

Theorem 4.3.1. *If L is chosen so that equation (4.65) holds, and $c_l(r)$ are defined by the formula*

$$c_l(r) = a_l f_l j_l(kr), \qquad (4.72)$$

where a_l is defined in equation (4.71), then the function

$$h_L(y) := \sum_{l=0}^{L} c_l(r) Y_l(y^0), \quad r = |y|, \quad y^0 := \frac{y}{r}, \qquad (4.73)$$

satisfies the inequality (4.55).

Let us give a formula for a potential q such that qu_q approximates h_L, defined in equation (4.73), with a desired accuracy. This h_L we denote by h in what follows. Using the h, defined in equation (4.73), calculate the potential q by formula (4.33). If this q belongs to $L^2(B)$, then the underdetermined inverse problem (IP) is solved.

If $q \notin L^2(B)$, then compute h_δ in place of h by formula (4.36) and q_δ by formula (4.37). This q_δ is bounded, and the corresponding to q_δ scattering amplitude $\frac{1}{4\pi} \int_B e^{-ik\beta \cdot y} h_\delta(y) dy$ satisfies inequality (4.55).

One can suggest a different computational method for finding $h(y)$ if $f(\beta)$ is given. For example, choose a basis $\{\phi_j\}$ in $L^2(B)$ and look for h_n:

$$h_n(x) = \sum_{j=1}^{n} a_j^{(n)} \phi_j(x), \qquad (4.74)$$

where $a_j^{(n)}$ are some coefficients which are found from the following minimization problem:

$$\left\| f(\beta) - \sum_{j=1}^{n} a_j^{(n)} g_j(\beta) \right\|_{L^2(S^2)} = \min, \qquad (4.75)$$

where

$$g_j := g_j(\beta) := -\frac{1}{4\pi} \int_B e^{-ik\beta \cdot y} \phi_j(y) dy. \qquad (4.76)$$

A necessary condition for the coefficients $a_j^{(n)}$ to yield minimum in equation (4.75) is a linear algebraic system

$$\sum_{j=1}^{n} (g_j, g_m) a_j^{(n)} = (f, g_m), \quad 1 \leqslant m \leqslant n. \qquad (4.77)$$

We assume that the set $\{g_j(\beta)\}$, where $g_j(\beta)$ are defined in equation (4.76), is a linearly independent set. This is a nontrivial assumption since the equation

$$\int_B e^{-ik\beta \cdot y}\phi(y)dy = 0, \quad \forall \beta \in S^2, \tag{4.78}$$

where $k = \text{const} > 0$ is fixed, has many nontrivial solutions. Therefore, the suggested method, which seems simple, requires checking that the set $\{g_j\}_{j=1}^n$ is linearly independent.

4.4 Open problems

Our construction of q given the underdetermined scattering data did not resolve the following problems, which are currently open:
1. Can our algorithm for finding q yield a real-valued q?
2. Can it yield a q such that $\text{Im } q \leqslant 0$?

Both questions are of interest because we have proved uniqueness of the scattering solutions assuming $\text{Im } q \leqslant 0$.

4.5 Summary of the results

In this chapter the following underdetermined inverse problem (IP) is investigated and solved:

(IP): Given an arbitrary $f \in L^2(S^2)$ and an arbitrary small fixed $\varepsilon > 0$, can one find $q \in L^2(\Omega)$, where $\Omega \in \mathbb{R}^3$ is an arbitrary fixed bounded domain, such that inequality (4.5) holds, where

$$A_q(\beta) = -\frac{1}{4\pi}\int_\Omega e^{-ik\beta \cdot y}q(y)u(y, \alpha, k)dy, \tag{4.79}$$

$k > 0$ and $\alpha \in S^2$ are fixed?

It is proved that the IP has a solution and there are infinitely many solutions to this IP. Some formulas for a solution to IP are given.

References

[1] Ramm A G 2006 *Inverse Problems* (New York: Springer)
[2] Ramm A G 1987 Completeness of the products of solutions to PDE and uniqueness theorems in inverse scattering *Inverse Problems* 3 L77–82
[3] Ramm A G 1988 Recovery of the potential from fixed energy scattering data *Inverse Problems* 4 877–86
    Ramm A G 1989 Recovery of the potential from fixed energy scattering data *Inverse Problems* 5 255
[4] Ramm A G 1988 Multidimensional inverse problems and completeness of the products of solutions to PDE *J. Math. Anal. Appl.* 134 211–53
    Ramm A G 1989 Multidimensional inverse problems and completeness of the products of solutions to PDE *J. Math. Anal. Appl.* 139 302

[5] Ramm A G 1988 Uniqueness theorems for multidimensional inverse problems with unbounded coefficients *J. Math. Anal. Appl.* **136** 568–74

[6] Ramm A G 1989 Multidimensional inverse scattering problems and completeness of the products of solutions to homogeneous PDE *Z. Angew. Math. Mech.* **69** T13–22

[7] Ramm A G 1992 *Multidimensional Inverse Scattering Problems* (New York: Longman/ Wiley)

[8] Ramm A G 1994 *Multidimensional Inverse Scattering Problems* (Moscow: Mir)

[9] Ramm A G 2002 Stability of solutions to inverse scattering problems with fixed-energy data *Milan J. Math.* **70** 97–161

[10] Ramm A G 2009 Inverse scattering with non-overdetermined data *Phys. Lett.* A **373** 2988–91

[11] Ramm A G 2010 Uniqueness theorem for inverse scattering problem with non-overdetermined data *J. Phys. A: Math. Theor.* **43** 112001

[12] Ramm A G 2010 Uniqueness of the solution to inverse scattering problem with back-scattering data *Eurasian Math. J.* **1** 97–111

[13] Ramm A G 2011 Uniqueness of the solution to inverse scattering problem with scattering data at a fixed direction of the incident wave *J. Math. Phys.* **52** 123506

[14] Ramm A G 1986 *Scattering by Obstacles* (Dordrecht: D. Reidel) pp 1–442

[15] Lebedev N 1972 *Special Functions and their Applications* (New York: Dover)

IOP Publishing

Creating Materials with a Desired Refraction Coefficient
(Second Edition)

Alexander G Ramm

Chapter 5

On non-over-determined inverse problems

5.1 Introduction

Sections 5.1 and 5.2 are based on papers [1] and [2]. Section 5.3 is based on the paper [3]. A detailed presentation of this theory is given in [4].

The inverse scattering problem consists of finding the unknown surface S of a body D from the scattering data. These data are the values of the scattering amplitude $A(\beta, \alpha, k)$ at some values of β, α, k. The inverse scattering problem is a major theoretical problem of physics which has huge practical significance. The non-over-determined scattering data for obstacle scattering are the values $A(\beta) = A(\beta, \alpha_0, k_0)$ for a fixed $\alpha_0 \in S^2$ and a fixed $k = k_0 > 0$.

The principal significance of these data is clear: these are the minimal data allowing unique recovery of S. In general, the non-over-determined scattering data are the data depending on the same number of variables as the unknown object. In obstacle scattering the unknown object is the surface of the obstacle (body), so it is a two-dimensional object. Therefore, the non-over-determined scattering data must be a function of two variables. Such are the data $A(\beta)$, $\forall \beta \in S^2$.

The basic uniqueness theorems for solving the inverse scattering problem with non-over-determined scattering data belong to the author, see [1] and [5]. These results were not known for many decades. The history of the inverse scattering problem for obstacles (bodies) one finds in [6], see also [7] and [8].

The aim of this paper is to give a short proof of the following uniqueness result:

Theorem 5.1.1. *The data $A(\beta)$ known for all $\beta \in S^2$ determine the surface S of the obstacle uniquely.*

In section 5.2 a proof of theorem 5.1.1 is given and various remarks are made.

doi:10.1088/978-0-7503-3391-7ch5

Let us introduce some notations: D is a bounded connected C^2-smooth body (obstacle) with boundary S, $D' := \mathbb{R}^3 \backslash D$. If two bodies D_j, $j = 1, 2$, are considered, then $D_{12} := D_1 \cup D_2$, S_{12} is the boundary of D_{12}, $\mathcal{D} := D^{12} := D_1 \cap D_2$, and S is the boundary of $\mathcal{D}$.

The scattering solution is the unique solution to the following problem:

$$(\nabla^2 + k^2)u = 0 \quad \text{in} \quad D', \quad u|_S = 0, \tag{5.1}$$

$$u = e^{ik\alpha \cdot x} + v := u_0 + v, \quad u_0 := e^{ik\alpha \cdot x}, \tag{5.2}$$

v is the scattered field satisfying the radiation condition,

$$v = A(\beta, \alpha, k)\frac{e^{ikr}}{r} + O\left(\frac{1}{r^2}\right), \quad r := |x| \to \infty, \quad \frac{x}{r} = \beta, \tag{5.3}$$

where $\alpha, \beta \in S^2$, S^2 is the unit sphere, β is the direction of the scattered wave, α is the direction of the incident wave, $k = k_0 > 0$ is a fixed constant. The function $A(\beta, \alpha, k)$, the scattering amplitude, can be measured experimentally. Let us call it the scattering data. As we mentioned above, we assume in this paper that $A(\beta) = A(\beta, \alpha_0, k_0)$ is known for all $\beta \in S^2$, $\alpha = \alpha_0$ and $k = k_0$ are fixed.

5.2 Proof of theorem 5.1.1

Suppose the conclusion of theorem 5.1.1 is not true. Then there are three cases to consider:

(a) $D^{12} = \emptyset$,

(b) $D_1 \subset D_2$, and

(c) D^{12} is a proper part of D_1 and of D_2.

Let us consider these cases.

Case (a). In this case the scattering solutions u_j, $j = 1, 2$, can be uniquely analytically continued as solutions to equation (5.1) to the whole space $\mathbb{R}^3$ by setting $u_1 = u_2$ in D_1 and $u_2 = u_1$ in D_2. Note that if $A_1(\beta) = A_2(\beta)$ then $u_1 = u_2$ in D'_{12} by lemma 1 in [6, p 25]. This lemma says that if a function u satisfies equation (5.1) near infinity and

$$\lim_{R \to \infty} \int_{|x|=R} |u|^2 \, ds = 0, \tag{5.4}$$

then $u = 0$ everywhere where u satisfies equation (5.1). Thus, $u := u_1 = u_2$ is defined in $\mathbb{R}^3$, and u is the scattering solution, that is, $u = u_0 + v$, where the scattered field v satisfies the radiation condition (5.3), $u_0 := e^{ik\alpha_0 \cdot x}$, and $u = 0$ on S_1 and on S_2. Therefore, v satisfies equation (5.1) in $\mathbb{R}^3$ and the radiation condition.

Claim 5.2.1. If v solves equation (5.1) in $\mathbb{R}^3$ and satisfies the radiation condition at infinity, it must vanish in $\mathbb{R}^3$.

Proof of the claim. For convenience of the reader let us prove the above claim following [6]. Let $\bar{v}$ denote the complex conjugate of v. Multiply equation (5.1) for v by $\bar{v}$, integrate over the ball of radius R, then integrate by parts taking into account that v and $\bar{v}$ solve equation (5.1) in $\mathbb{R}^3$. The result is the relation: equation (5.4) for v. Since v solves equation (5.1) in $\mathbb{R}^3$ and equation (5.4) holds for v, one concludes that $v = 0$ in $\mathbb{R}^3$. □

If $v = 0$ in $\mathbb{R}^3$ then one has a contradiction since then $u = u_0$ and $u_0 \neq 0$ on S_j, $j = 1, 2$, because $|u_0| = 1$. So, **case (a)** cannot occur if $A_1(\beta) = A_2(\beta)$ for all $\beta \in S^2$. □

Case (b). In this case the function u_2 can be uniquely and analytically continued to D_1' as a solution to equation (5.1) by setting $u_2 = u_1 := u$ in D_1'. By Green's formula one gets

$$u_j(x) = u_0(x) - \int_{S_j} g(x, s) u_{jN_s} ds, \qquad g(x, y) := \frac{e^{ik|x-y|}}{4\pi|x - y|}, \tag{5.5}$$

where u_{N_s} is the normal derivative of u at the point s, $j = 1, 2$, formula (5.5) holds for $x \in D_j'$, and $u = u_2 = u_1$ in D_1', $u = 0$ on S_j, $j = 1, 2 \ldots$ Also, Green's formula yields

$$u(x) = \int_{S_2} g(x, s) u_{N_s} ds - \int_{S_1} g(x, s) u_{N_s} ds, \quad x \in D_2 \backslash D_1. \tag{5.6}$$

We derive a contradiction from equations (5.5) and (5.6). Indeed, from these equations one gets

$$T^+ := \int_{S_2} g(x, s) u_{N_s} ds = u + u_0 - u = u_0, \quad x \in D_2 \backslash D_1, \tag{5.7}$$

where we have used formula (5.5) with $j = 1$ to get $\int_{S_1} g(x, s) u_{N_s} ds = u_0 - u$ in $D_2 \backslash D_1$ and took into account that $u = u_1 = u_2$. On the other hand, equation (5.2) with $j = 2$ yields

$$T^- := \int_{S_2} g(x, s) u_{N_s} ds = u_0 - u, \quad x \in D_2'. \tag{5.8}$$

The left side of equation (5.8) admits a unique continuation into $D_2 \backslash D_1$ as T^+ since u and u_0 admit such a continuation. Comparing this continuation with equation (5.7) one concludes that $u = 0$ in $D_2 \backslash D_1$. Therefore, $u = 0$ in D_1' because of the unique continuation property of the solution u to the elliptic equation (5.1) in D_1'.

If $u = 0$ in D_1' then we have a contradiction since $\lim_{|x| \to \infty} |u| = 1$. This contradiction proves that **case (b)** cannot occur. □

Case (c). The proof in this case is essentially the proof used in **case (b)**.

If $A_1(\beta) = A_2(\beta)$ then $u_1 = u_2$ in D_{12}. The function u_1 admits a unique continuation as a solution to equation (5.1) in $\mathcal{D}'$ by the formula $u_1 = u_2$ in $D_1 \backslash D$, and u_2 admits such a continuation by the formula $u_2 = u_1$ in $D_2 \backslash D$. Therefore, the function

$u = u_1 = u_2$ in $\mathcal{D}'$ satisfies equation (5.1), $u = 0$ on S_{12} and on S, and $u = u_0 + v$, where v satisfies the radiation condition. Now the contradiction which shows that **case (c)** cannot occur unless $D_1 = D_2$ is derived as in **case (b)**. The role of S_2 is played by S_{12} and the role of S_1 is played by S, the boundary of D^{12}. The applicability of the Green's formula is justified in **case (c)** in remark 5.2.2 below. $\qquad\square$

Theorem 5.1.2 *By remark 5.2.3 (see below) the conclusion of theorem 5.1.1 remains valid if the non-over-determined scattering data $A(\beta)$ is known on an open subset of S^2, however small.* $\qquad\square$

Remark 5.2.1. *One can use the proof of theorem 5.1.1 for proving the uniqueness theorem for the Neumann and the Robin boundary conditions without essential changes. The proof allows one to determine uniquely the boundary condition on S. Indeed, if S is uniquely determined then u is uniquely determined in D'. One can calculate u and u_N on S. If $u|_S = 0$ then the Dirichlet boundary condition holds. If $u_N|_S = 0$ then the Neumann boundary condition holds. If $u_N/u := \zeta(s)$ then the Robin (impedance) boundary condition holds. We assume that Im $\zeta(s) \leqslant 0$. This guarantees uniqueness of the solution to the scattering problem with impedance boundary condition, see [6]. Many-body scattering theory for small impedance bodies of arbitrary shapes is developed in [9].*

Remark 5.2.2. *In case (c) the smooth surfaces S_1 and S_2 intersect. Intersection of two smooth surfaces may be not smooth. However, one can use Green's formula in D'_{12} and in $\mathcal{D}'$. Indeed, u_j are smooth and bounded up to the boundary S_{12} together with its derivatives. The non-smooth parts of S_{12} are of two-dimensional Lebesgue measure zero. They can be covered by smooth patches S_ε so that the resulting surface will be smooth and will converge to S_{12} as $\varepsilon \to 0$. The input in Green's formula from S_ε tends to zero as $\varepsilon \to 0$ because u_j and its derivatives are bounded on S_{12} and in its neighborhood. This justifies the usage of Green's formula.*

Remark 5.2.3. *It is proved in [6, p 62], that if $A(\beta)$ is known on an open subset of S^2 then it is uniquely defined on all of S^2. Therefore, the conclusion of theorem 5.1.1 remains valid if the data $A(\beta)$ is known on an open subset of S^2 rather than on all of S^2.*

5.3 A numerical method

The inverse scattering problem consists of finding the unknown potential $q(x)$ from the scattering data. These data are the values of the scattering amplitude $A(\beta, \alpha, k)$ at some values of β, α, k. The inverse scattering problem is a major theoretical problem of physics which has huge practical significance.

The basic uniqueness theorem for solving the inverse scattering problem with non-over-determined scattering data belongs to the author, [5]. This result was not known for decades. There were no results on numerical methods for solving the inverse scattering problem with non-over-determined data.

The inverse scattering problem is highly non-linear because the scattering amplitude depends non-linearly on the potential. Therefore, it is remarkable that the inversion procedure proposed in this paper is linear: it is reduced to numerical solution of a linear algebraic system, see system (5.16) below. No such results were known. The inverse scattering problem is not solved by a parameter fitting procedure. The numerical method is based on the author's uniqueness theorem [5].

The scattering solution is the unique solution to the following problem:

$$(\nabla^2 + k^2 - q(x))u = 0 \quad \text{in} \quad \mathbb{R}^3, \tag{5.9}$$

$$u = e^{ik\alpha \cdot x} + v, \tag{5.10}$$

where v is the scattered field satisfying the radiation condition,

$$v = A(\beta, \alpha, k)\frac{e^{ikr}}{r} + o\left(\frac{1}{r}\right), \quad r := |x| \to \infty, \quad \frac{x}{r} = \beta. \tag{5.11}$$

where $\alpha, \beta \in S^2$, S^2 is the unit sphere, β is the direction of the scattered wave, α is the direction of the incident wave, $k^2 > 0$ is energy, $k > 0$ is a constant. The function $A(\beta, \alpha, k)$, the scattering amplitude, can be measured experimentally. Let us call it the scattering data.

We assume throughout that q is a real-valued compactly supported function with support D, $q = 0$ for $x \notin D$, $D = \{x: \max_j |x_j| \leqslant R\}$, and q is C^1-smooth. The set of such q let us call Q.

The assumption that q is compactly supported is natural and even necessary when one solves ISP with noisy data, see [8] for details.

It is known that the solution to the scattering problem (5.9)–(5.11) does exist and is unique (see, e.g. [7]).

The inverse scattering problem (IP) consists of finding $q \in Q$ from the scattering data.

It was first proved by A G Ramm [7, 10] that $q \in Q$ is uniquely determined by the scattering data $A(\beta, \alpha, k_0)$ known for a fixed $k = k_0 > 0$ and all $\beta \in S^2_\beta$ and all $\alpha \in S^2_\alpha$, where S^2_β is an open subset of S^2.

A G Ramm gave a method for solving the inverse scattering problem with fixed-energy data and obtained an error estimate for the solution for exact data and also for noisy data, see [7].

The goal of this section is to give a numerical method for solving the inverse scattering problem with non-over-determined scattering data. The non-over-determined scattering data are the data that depend on the same number of variables as the potential, that is, on three variables. We assume that these data are the values of $A(\beta, k) := A(\beta, \alpha_0, k)$ known for all $\beta \in S^2_\beta$, for all $k \in (a, b), 0 \leqslant a < b$, and a fixed $\alpha_0 \in S^2$.

Our method for solving this inverse scattering problem is described in section 5.2. This problem is reduced to solving linear algebraic system which is very ill-conditioned.

Therefore, numerically one should use the DSM (dynamical systems method), [11, 12] a stable method for solving linear algebraic system (5.17) (or other stable methods for numerical solution of ill-conditioned linear algebraic systems, see [11]). Stable solution of equation (5.17) is the main numerical difficulty of our method. This method is not a parameter fitting method, which is a big advantage of the method. There were no numerical methods for solving the inverse scattering problem with non-over-determined data. The theoretical basis for our paper is the uniqueness theorem proved by the author in [5].

Let us describe our inversion method.

The scattering problem is equivalent to the standard integral equation (see, e.g. [7]):

$$u = e^{ik\alpha_0 x} - \int_D g(x, y, k)q(y)u(y, \alpha_0, k)dy, \qquad g(x, y, k) := \frac{e^{ik|x-y|}}{4\pi|x - y|}, \quad (5.12)$$

where the integral is taken over the support of $q(x)$ and the dependence on the fixed vector α_0 is dropped in what follows. Define

$$h := q(x)u(x, k). \tag{5.13}$$

Equation (5.12) implies the following equation for h:

$$h = q(x)e^{ik\alpha_0 x} - q(x) \int_D g(x, y, k)h(y, k)dy. \tag{5.14}$$

From equation (5.12) one derives the following exact formula for the scattering amplitude:

$$-4\pi A(\beta, k) = \int_D e^{-ik\beta \cdot y}h(y, k)dy, \tag{5.15}$$

where $\beta \in S^2$ and $k \in (a, b)$. Recall that we write $A(\beta, k)$ for $A(\beta, \alpha_0, k)$ and $h(x, k)$ for $h(x, \alpha_0, k)$.

If $A(\beta, k)$ is known, then equation (5.15) is a linear integral equation of the first kind with respect to the unknown $h(y, k)$. If h is found, then q can be found by formula (5.17) below.

Let us partition the support of q into a union of P small cubes Δ_p. Choose a point $y_p \in \Delta_p$, $1 \leqslant p \leqslant P$, in each of the small cubes. Denote by Δ the volume of each small cube. Choose P different points $k_m \in (a, b)$, $1 \leqslant m \leqslant P$. Denote $h_{pm} := h(y_p, k_m)$. Choose P different vectors $\beta_j \in S_\beta^2, 1 \leqslant j \leqslant P$. Discretize equation (5.15):

$$-4\pi A(\beta_j, k_m) = \sum_{p=1}^{P} e^{-ik_m \beta_j \cdot y_p}h_{pm}\Delta, \qquad 1 \leqslant j, m \leqslant P, \tag{5.16}$$

where Δ is the element of the volume of the support of q.

Equation (5.16) is a linear algebraic system of P^2 equations for P^2 unknowns h_{pm}, $1 \leqslant p, m, j \leqslant P$. If this system is solved numerically, then equation (5.14) yields the values $q(x_p)$ of the unknown potential:

$$q(x_p) = h_{pm} \left[e^{ik_m\alpha_0 x_p} - \sum_{p'=1,p'\neq p}^{P} g(x_p, y_{p'}, k_m) h_{p'm} \Delta \right]^{-1}, \qquad (5.17)$$

where $1 \leqslant p \leqslant P$, and the right side of equation (5.17) should not depend on m or j.

Although the right side of equation (5.17) does not depend on j explicitly, it does depend on j implicitly since there is a dependence on j in equation (5.16), so that the solution h_{pm} of equation (5.16) does depend on j.

The independence of $q(x)$ and, therefore, the right side of equation (5.17) on m and j is an important requirement in the numerical solution of the inverse scattering problem, a compatibility condition for the data. This requirement is automatically satisfied for the limiting integral equation formula:

$$q(x) = h(x, k) \left[e^{ik\alpha_0 x} - \int_D g(x, y, k) h(y, k) dy \right]^{-1}, \qquad (5.18)$$

which follows from equation (5.14).

The values $q(y_p)$ essentially determine the C^1-smooth potential q if the distance between the neighboring points y_p is sufficiently small.

The linear algebraic system (5.16) is very ill-conditioned because it comes from an integral equation of the first kind with an analytic kernel. From the author's uniqueness theorem (see [5]) it follows that *the non-over-determined scattering data $A(\beta, k)$ determine uniquely the potential $q \in Q$.*

Thus, one expects that the proposed method can be numerically efficient if the linear algebraic system (5.16) is solved stably. Theoretical methods for stable solution of ill-conditioned linear algebraic systems are developed in [11] and in [12] one finds many numerical examples of such solutions.

There were no numerical methods for solving the inverse scattering problem with non-over-determined data, as far as the author knows.

One can choose β_j and k_m so that the determinant of the linear algebraic system (5.16) is not equal to zero, so that the system is uniquely solvable. This does not eliminate the essential difficulties in numerical solution of the inverse scattering problem caused by the numerical difficulties in solving severely ill-conditioned linear algebraic systems. Therefore, it is essential to apply the DSM (see [11, 12]) which proved to be an effective tool for solving ill-conditioned linear algebraic systems. In examples considered in [12] the DSM was more effective for solving ill-posed problems than the variational regularization method. One of the advantages of the DSM over the variational regularization is the automatic calculation of the stopping rule (without solving a nonlinear equation for finding the regularization parameter).

In conclusion let us prove the following lemma.

Lemma 5.3.1. *There exist* $\beta_j \in S_\beta^2$ *and* $k_m \in (a, b)$, $1 \leqslant j$, $m \leqslant P$, *such that*

$$\det(e^{-ik_m\beta_j \cdot y_p}) \neq 0.$$

In this lemma the matrix depends on m, p. The index j enters as a parameter, $1 \leqslant m$, $p \leqslant P$, $1 \leqslant j \leqslant P$.

Proof of lemma 5.3.1. Let $\beta_j \in S_\beta^2$ be arbitrarily fixed, $y_p \neq y_{p'}$ if $p \neq p'$, and $b_{pj} := \beta_j \cdot y_p \neq b_{p'j}$ if $p \neq p'$. Let us prove that there are $k_m \in (a, b)$, $1 \leqslant m \leqslant P$, such that $\det(e^{-ik_m b_{pj}}) \neq 0$. Assume the contrary. Then $\det(e^{-ik_m b_{pj}}) = 0$ for any choice of k_m. The function $e^{-ikb_{pj}}$ is analytic (entire) with respect to k. Therefore, if the above determinant vanishes for all $k_1 = k$, then it vanishes identically with respect to k, so that the set of functions $\{e^{ikb_{pj}}\}_{p=1}^P$ is linearly dependent. This is a contradiction since the above set is linearly independent under our assumption, namely the assumption that $b_{pj} \neq b_{p'j}$ if $p \neq p'$. Indeed, if c_p are constants and $\sum_{p=1}^P c_p e^{-ikb_{pj}} = 0$ for all $k \in (a, b)$, then, by analyticity, $\sum_{p=1}^P c_p e^{sb_{pj}} = 0$ for all $s \in \mathbb{R}$. Since all numbers b_{pj} are different and real-valued, they can be ordered. Let us assume without loss of generality that $b_1 > b_2 > \cdots > b_P$, where $b_p := b_{pj}$. Then, the relation $c_1 + \sum_{p=2}^P c_p e^{-s(b_1 - b_p)} = 0$ for $s \to +\infty$ yields $c_1 = 0$. Similarly, one proves that $c_p = 0$ for all p. This contradicts to the linear dependence of the system $\{e^{ikb_{pj}}\}_{p=1}^P$. Lemma 5.3.1 is proved. $\qquad\square$

5.4 Summary of the results

In this chapter the obstacle inverse scattering problem with fixed-energy data is studied. The basic result is theorem 5.1.1.

In section 5.3 a numerical approach to this problem is discussed.

References

[1] Ramm A G 2016 Uniqueness of the solution to inverse obstacle scattering with non-over-determined data *Appl. Math. Lett.* **58** 81–6
[2] Ramm A G 2018 A uniqueness theorem for inverse scattering problem with non-over-determined data *J. Phys. A: Math. Theor.* **43** 112001
[3] Ramm A G 2017 A numerical method for solving 3D inverse scattering problem with non-over-determined data *J. Pure Appl. Math.* **1** 1–3
[4] Ramm A G 2019 *Inverse Obstacle Scattering with Non-Over-Determined Scattering Data* (San Rafael, CA: Morgan & Claypool)
[5] Ramm A G 2011 Uniqueness of the solution to inverse scattering problem with scattering data at a fixed direction of the incident wave *J. Math. Phys.* **52** 123506
[6] Ramm A G 1986 *Scattering by Obstacles* (Dordrecht: D. Reidel) pp 1–442
[7] Ramm A G 2005 *Inverse Problems* (New York: Springer)
[8] Ramm A G 2017 Finding a method for producing small impedance particles with prescribed boundary impedance is important *J. Phys. Res. Appl.* **1** 1–3

[9] Ramm A G 2013 *Scattering of Acoustic and Electromagnetic Waves by Small Bodies of Arbitrary Shapes. Applications to Creating New Engineered Materials* (New York: Momentum)

[10] Ramm A G 1988 Recovery of the potential from fixed energy scattering data *Inverse Problems* **4** 877–86

Ramm A G 1989 Recovery of the potential from fixed energy scattering data *Inverse Problems* **5** 255

[11] Ramm A G 2007 *Dynamical Systems Method for Solving Operator Equations* (Amsterdam: Elsevier)

[12] Ramm A G and Hoang N S 2012 *Dynamical Systems Method and Applications. Theoretical Developments and Numerical Examples* (Hoboken, NJ: Wiley)

IOP Publishing

Creating Materials with a Desired Refraction Coefficient
(Second Edition)

Alexander G Ramm

Chapter 6

Experimental verification of the method for creating materials

6.1 Moving the refraction coefficient in the desired direction

In chapter 3 a recipe is given for creating materials with a desired refraction coefficient. This recipe requires embedding many small particles with prescribed boundary impedances into a given material.

Suppose that we just wish to create a material with a smaller refraction coefficient than that of the given material. This is an easier problem. It is of practical importance. Indeed, one is often interested in creating materials with a very small refraction coefficient or meta-materials with a negative refraction coefficient.

How does one create materials with a smaller refraction coefficient $n(x)$ given a material with the refraction coefficient $n_0(x)$?

One uses the recipe described in section 3.2. Let us take $N(x) = N =$ const. This means that the number of particles in any subset Δ of the domain D depends only on the volume of Δ but not on the position of Δ in D. If $n(x) < n_0(x)$, then the function

$$p(x) = k^2[n_0^2(x) - n^2(x)], \qquad (6.1)$$

see formula (3.4), is positive. Since N is constant, formula (3.5) yields

$$p(x) = 4\pi N h(x). \qquad (6.2)$$

So, see formula (3.7),

$$h(x) = \frac{p(x)}{4\pi N}. \qquad (6.3)$$

doi:10.1088/978-0-7503-3391-7ch6

We have assumed that

$$\text{Im } n_0^2(x) = 0, \quad \text{Im } n^2(x) = 0, \tag{6.4}$$

so $p_1(x)$ in formula (3.7) equals $p(x)$ and $p_2(x) = 0$.

Choose $\kappa = \frac{1}{2}$. Then, by formula (2.130) one has

$$\zeta_m = \frac{h(x_m)}{\sqrt{a}}, \quad 1 \leqslant m \leqslant M, \quad \text{Im } h(x) = 0, \tag{6.5}$$

where a is the radius of a particle and ζ_m is its boundary impedance.

Distribute the spherical particles of radius a in the domain D so that in any subdomain $\Delta \subset D$ there are

$$\mathcal{N}(\Delta) = \frac{N}{a^{3/2}} |\Delta|, \tag{6.6}$$

particles, where $|\Delta|$ is the volume of Δ.

Assume that the distance d between neighboring particles is much larger than a:

$$d \gg a. \tag{6.7}$$

The approximate solution of the scattering problem (2.119)–(2.122) is given by formula (2.134). The limiting form of equation (2.134) under the assumptions (2.141) is the equation (2.145).

In this equation $c = 4\pi$ (because the surface area of the ball of radius a is equal to $ca^2 = 4\pi a^2$), $N(y) = N$, and the corresponding refraction coefficient is

$$n^2(x) = 1 - \frac{2\pi N}{k^2} h(x), \tag{6.8}$$

where we assumed for simplicity that $n_0^2(x) = 1$ in D, $n_0^2(x) = 0$ in $D' = \mathbb{R}^3 \backslash D$. The function $h(x)$ is such that

$$h(x_m) = a^{1/2} \zeta_m, \tag{6.9}$$

where x_m is the center of the mth ball D_m (particle) and ζ_m is the boundary impedance of D_m, $1 \leqslant m \leqslant M$, see formula (2.130).

The function $h(x)$ for $x \neq x_m$ can be constructed by an interpolation. The choice of the interpolation formula is not important.

We remind the reader that the choice of ζ_m (and, consequently, the functions $h(x)$ and $n^2(x)$) is in the hands of the experimentalist.

Formula (6.8) shows that $n^2(x) < 1 = n_0^2(x)$, as desired.

Formula (2.145) is the limiting form of the linear algebraic system (LAS) (2.144). For this LAS to approximate accurately equation (2.145) one needs many small particles (balls) to be embedded into D and conditions

$$a \ll d \ll b, \quad \lim_{a \to 0} \frac{a}{d(a)} = \lim_{a \to 0} \frac{d(a)}{b(a)} = 0, \tag{6.10}$$

to hold, where $b = b(a)$, $d = d(a)$, $b = \max_p \text{diam } \Delta_p$, Δ_p is a subset of D, $\Delta_p \cap \Delta_j \neq \varnothing$ if $p \neq j$, $\cup_{p=1}^P \Delta_p = D$.

Whether formula (6.8) is sufficiently accurate one can verify experimentally by taking particles of radius $\frac{a}{2}$ and checking if the resulting formula yields approximately the same $n^2(x)$ as formula corresponding to radius a.

The practical problems are:

How does one prepare small particles with the prescribed boundary impedance?

Do such particles exist in nature?

The author thinks that any particles such that the boundary conditions on their surfaces guarantee the uniqueness and existence of the solution to the Maxwell's boundary problem do exist in nature. The impedance boundary condition (2.2) under our assumption guarantees existence and uniqueness of the solution to the scattering problem. Therefore, the answer to the second question is: *yes*.

A physical argument proving this goes as follows. The particles with $\zeta = \infty$ (acoustically soft particles) and the particles with $\zeta = 0$ (acoustically hard particles) do exist in nature. Therefore, the particles with an intermediate value of ζ should also exist in nature.

6.2 The case of a bounded region

In this section we discuss the case when the material and the embedded small impedance particles are located not in a region of the infinite space $\mathbb{R}^3$ but in a bounded domain Ω with a smooth boundary Γ on which the Dirichlet boundary condition holds. The corresponding scattering problem under the simplifying assumptions used in section 6.1 is (compared with equations (2.1)–(2.4)):

$$(\Delta + k^2)u = 0 \quad \text{in} \quad D' := \Omega \backslash D, \tag{6.11}$$

$$u_{\bar{N}} = \zeta u \quad \text{on} \quad S, \quad \text{Im } \zeta < 0, \tag{6.12}$$

$$u|_\Gamma = 0, \tag{6.13}$$

$$u = u_0 + v, \quad u_0 = e^{ik\alpha \cdot x}, \quad \alpha \in S^2. \tag{6.14}$$

This is a boundary value problem with

$$S = \cup_{j=1}^M S_j, \quad D = \cup_{j=1}^M D_j, \quad S_j = \partial D_j.$$

Our *first main assumption* is:

$$k^2 \neq k_p^2, \tag{6.15}$$

where k_p^2 are the eigenvalues of the Dirichlet Laplacian in Ω:

$$(\Delta + k^2)w = 0 \quad \text{in} \quad \Omega, \quad w|_\Gamma = 0. \tag{6.16}$$

Our *second main assumption* is equation (6.10).

One proves the analogs of theorems 2.1.1 and 2.1.2 in which formula (2.10) is modified as follows

$$u = u_0 + \sum_{j=1}^{M} \int_{S_j} g(x, t)\sigma_j(t)dt + \int_{\Gamma} g(x, t)\sigma_0(t)dt. \qquad (6.17)$$

Theorem 6.2.1. *Problem (6.11)–(6.14) has at most one solution.*

Proof. It is sufficient to prove that $u = 0$ if $u_0 = 0$. As in section 2.1.1, we have:

$$0 = \int_{\Gamma} (\bar{u}u_N - u\bar{u}_N)ds - \int_{S} (\bar{u}u_N - u\bar{u}_N)ds. \qquad (6.18)$$

Since $u|_\Gamma = 0$, one has

$$\int_{S} (\zeta|u|^2 - \bar{\zeta}|u|^2)ds = 0. \qquad (6.19)$$

Since Im $\zeta < 0$, it follows from equation (6.19) that $u = 0$ on S. So, $u_{\bar{N}}|_S = 0$. This and the uniqueness of the solution to the Cauchy problem for elliptic equation (6.11) imply that $u = 0$.

Theorem 6.2.1 is proved. □

Theorem 6.2.2. *Problem (6.11)–(6.14) has a solution and this solution is unique.*

Proof. The uniqueness of the solution was proved in theorem 6.2.1. Let us prove the existence of the solution.

The function (6.17) will be the solution of the problem (6.11)–(6.14) if the boundary conditions (6.12) and (6.13) are satisfied. This yields $M + 1$ linear Fredholm-type integral equations for $M + 1$ unknown functions σ_j, $0 \leqslant j \leqslant M$. The homogeneous system of integral equations, corresponding to the assumption $u_0 = 0$, has only the trivial solution by theorem 6.2.1. Therefore, the solution to the system of integral equations with $u_0 \neq 0$ does exist and is unique.

The operator $T\sigma_0 = \int_{\Gamma} g(x, t)\sigma_0(t)dt$ is compact in $L^2(\Gamma)$ (if Γ is sufficiently smooth, as we assume). It is of Fredholm type as an operator from $L^2(\Gamma)$ to $H^1(\Gamma)$ if equation (6.15) holds. This means that if equation (6.15) holds then the null-space of T is trivial $(N(T) = 0)$, and the range of T is $H^1(\Gamma)$, $R(T) = H^1(\Gamma)$. Indeed, to check that $N(T) = 0$ assume that $T\sigma_0 = 0$. Then $w(x) = \int_{\Gamma} g(x, t)\sigma_0(t)dt$ solves the problem

$$(\nabla^2 + k^2)w = 0 \quad \text{in} \quad \Omega' := \mathbb{R}^3 \backslash \Omega, \quad w|_\Gamma = 0, \qquad (6.20)$$

and w satisfies the radiation condition. Therefore $w = 0$ in Ω'.

Furthermore,

$$(\nabla^2 + k^2)w = 0 \quad \text{in} \quad \Omega, \quad w|_\Gamma = 0. \tag{6.21}$$

Thus, $w = 0$ in Ω by assumption (6.15). Consequently, $\sigma_0 = w_N^+ - w_N^- = 0$. So, $N(T) = 0$.

Assume now that $f \in H^1(\Gamma)$ is arbitrary. Consider the problem:

$$(\nabla^2 + k^2)w = 0 \quad \text{in} \quad \Omega, \quad w|_\Gamma = f. \tag{6.22}$$

Problem (6.22) has a unique solution $w \in H^{3/2}(\Omega)$ due to assumption (6.15). By the trace theorem and the jump formula for the normal derivative of the simple layer potential one has $\sigma_0 \in L^2(\Gamma)$. Thus, the existence of the solution to problem (6.11)–(6.14) follows from the uniqueness of the solution to this problem.

Theorem 6.2.2 is proved. $\qquad\qquad\qquad\qquad\qquad\qquad\qquad\qquad\square$

If condition (6.10) holds then the effective field u can be written as

$$u(x) = u_0(x) + \int_\Gamma g(x, t)\sigma_0(t)dt + \sum_{j=1}^M g(x, x_j)Q_j, \quad Q_j = -4\pi h_j a^{2-\kappa} u_j, \tag{6.23}$$

see formula (2.133). The constants c_m in this formula are equal to 4π since all D_j are balls of radius a. Take $x = x_m$ in (6.23) and let $u_m = u(x_m)$, $u_0(x_m) = u_{0m}$. Then

$$u_m = u_{0m} + \int_\Gamma g(x_m, t)\sigma_0(t)dt - \sum_{j=1}^M 4\pi g_{mj} h_j u_j a^{2-\kappa}. \tag{6.24}$$

As in the derivation of equation (2.145), one gets

$$u(x) = u_0(x) + \int_\Gamma g(x, t)\sigma_0(t)dt - \sum_{j=1}^M 4\pi h_j a^{2-\kappa} u_j, \tag{6.25}$$

where

$$N(y) = \frac{\int_{\Delta_j} N(x)dx}{|\Delta_j|}, \quad y \in \Delta_j. \tag{6.26}$$

We assume that the domain $D \subset \Omega$ in which the small particles are distributed is a strictly inner subdomain of Ω. Taking the operator $\nabla^2 + k^2$ of the equation (6.25) yields

$$(\nabla^2 + k^2)u = 4\pi h(y)N(y)u(y) := q(y)u(y) \tag{6.27}$$

where we take into account that $(\nabla^2 + k^2)u_0 = 0$. Thus,

$$(\nabla^2 + k^2 - q(x))u = 0 \quad \text{in} \quad \Omega, \quad u|_\Gamma = 0, \ u = u_0 + v, \tag{6.28}$$

where $v = \int_\Gamma g(x, t)\sigma_0(t)dt - \int_D g(x, y)q(y)u(y)dy$. By Green's formula one gets

$$u(x) = u_0(x) + \int_\Gamma g(x, y)u_N(t)dt - \int_D g(x, y)q(y)u(y)dy, \quad u_0 = e^{ika\cdot x}. \quad (6.29)$$

Compare this with formula (6.25) and conclude:

$$\sigma_0(t) = u_N(t). \quad (6.30)$$

From equation (6.28) one gets

$$(\nabla^2 + k^2 n^2(x))u = 0, \quad n^2(x) = 1 - \frac{q(x)}{k^2}. \quad (6.31)$$

Therefore, the new refraction coefficient is smaller than the original one which we took (for simplicity) to be equal to 1. We have assumed that Im $h(x) < 0$, but one can choose Im $h(x)$ to be very small and neglect it, so that $g(x)$ can be considered positive from the practical point of view. The assumption Im $h(x) < 0$ guaranteed uniqueness of the solution to problem (6.11)–(6.14). In the next section we consider the case of acoustically soft small particles (in mechanics) or perfectly conducting particles (in electrodynamics). In this case $h(x) > 0$ and uniqueness of the solution to problem analogous to (6.11)–(6.14) is proved.

6.3 Embedding acoustically soft particles

The acoustically soft particles are the particles D_j on the boundary of which the Dirichlet boundary condition holds:

$$u|_{S\cup\Gamma} = 0, \quad S = \bigcup_{j=1}^M S_j, \quad (6.32)$$

instead of the impedance boundary condition (6.12). So the boundary problem now is equations (6.11), (6.32), (6.13), (6.14).

Analogs of theorems 6.2.1 and 6.2.2 hold, see [1].

As in section 2.2.2 one derives the equation

$$u(x) = u_0(x) + \int_\Gamma g(x, t)\sigma_0(t)dt - 4\pi \int_D g(x, y)N(y)\,u(y)dy, \quad (6.33)$$

where $u_0 = e^{ika\cdot x}$, $\sigma_0(t) = u_N(t)$ and

$$(\nabla^2 + k^2 - q(x))u = 0, \quad q = 4\pi N(x). \quad (6.34)$$

By formula (2.129) $N(x) = \frac{a\mathcal{N}(\Delta)}{|\Delta|}$ for small Δ, $|\Delta|$ is the volume of Δ. Assume (for simplicity only) that $N(x)$ does not depend on x in D. Then

$$\mathcal{N}(\Delta) = \rho|\Delta|, \quad \rho = \text{const}, \quad N = a\rho, \quad q(x) = 4\pi a\rho,$$

and

$$n^2(x) = n^2 = 1 - \frac{4\pi a\rho}{k^2}. \quad (6.35)$$

Thus, $n^2 < 1$. Therefore, if one embeds many small acoustically soft balls of radius a in the domain $D \subset \Omega$, so that assumptions (6.10) and (6.15) hold then one gets the new material with refraction coefficient which is given approximately by formula (6.35) in D.

6.4 Summary of the results

In this chapter a method for experimental verification of the theory for creating materials with a desired refraction coefficient is proposed. In particular, it is shown how to move the refraction coefficient in a desired direction, for example, how to make it smaller.

In section 6.2 the theory is developed for the case when the material is located in a bounded region on the boundary of which the Dirichlet boundary condition is imposed. The uniqueness and existence theorems are theorems 6.2.1 and 6.2.2.

In section 6.3 the embedding of small acoustically soft particles, rather than the impedance particles, is discussed. The material of this chapter is new. It is based on the theory developed in [1, 2].

References

[1] Ramm A G 2013 *Scattering of Acoustic and Electromagnetic Waves by Small Bodies of Arbitrary Shapes. Applications to Creating New Engineered Materials* (New York: Momentum)
[2] Ramm A G 2017 *Creating Materials with a Desired Refraction Coefficient* (IOP Concise Physics) (San Rafael, CA: Morgan & Claypool)

Chapter 7

A symmetry property in harmonic analysis

Many symmetry properties discovered by the author in scattering theory and for solutions to the Helmholtz equation are presented in [1]. Among these were the properties that allowed the author to solve the Pompeiu problem and to prove the refined Schiffer's conjecture, see [1]. In this chapter the author presents, probably, the first result of non-trivial symmetry problem in harmonic analysis.

Let us formulate the results.

Theorem 7.0.1. *Let* $u(k, \alpha) = \int_D e^{ik\alpha \cdot x} dx$, *where* $D \subset \mathbb{R}^3$ *is a bounded connected domain. If* u *has a spherical set of zeros, that is,* $u(k_0, \alpha) = 0$ *for all* $\alpha \in S^2$ *and a fixed* $k_0 > 0$, *then* D *is a ball.*

Theorem 7.0.2. *Let* $v(k, \alpha) = \int_S e^{ik\alpha \cdot s} ds$. *If* v *has a spherical set of zeros, that is,* $v(k_0, \alpha) = 0$ *for all* $\alpha \in S^2$ *and a fixed* $k_0 > 0$, *then* S *is a sphere.*

Proof of theorem 7.0.1. Consider the function

$$w(x) = \int_D g(x, y) dy, \quad g(x, y) := \frac{e^{ik_0|x-y|}}{4\pi|x - y|}. \tag{7.1}$$

Since D is bounded, one has

$$w(x) = \frac{e^{ik_0|x|}}{4\pi|x|} \int_D e^{ik_0\alpha \cdot y} dy + O\left(\frac{1}{|x|^2}\right) = O\left(\frac{1}{|x|^2}\right), \quad |x| \to \infty, \quad \frac{x}{|x|} = -\alpha, \tag{7.2}$$

where we have used *the assumption*

$$\int_D e^{ik_0\alpha \cdot y} dy = 0, \quad \forall \alpha \in S^2. \tag{7.3}$$

doi:10.1088/978-0-7503-3391-7ch7

One has

$$(\nabla^2 + k_0^2)w = -1 \quad \text{in} \quad D, \tag{7.4}$$

$$(\nabla^2 + k_0^2)w = 0 \quad \text{in} \quad D' := \mathbb{R}^3 \backslash D. \tag{7.5}$$

Condition (7.2) and equation (7.5) imply

$$w = 0 \quad \text{in} \quad D'. \tag{7.6}$$

This follows from lemma 1.2.1 on p 30 of [2]. This lemma says that any solution to equation (7.5) which is $o(\frac{1}{|x|})$ as $|x| \to \infty$ vanishes in D'. If $w = 0$ in D', then

$$w|_S = 0, \quad w_N^-|_S = 0, \quad S = \partial D, \tag{7.7}$$

where w_N^- is the limiting value of the normal derivative on S from outside, i.e., from D'. The potential w in equation (7.1) is continuous with its first derivatives across S. Therefore equation (7.4) and the boundary conditions

$$w|_S = 0, \quad w_N|_S = 0, \tag{7.8}$$

hold.

Let

$$w = \psi - \frac{1}{k_0^2}. \tag{7.9}$$

Then equation (7.4) yields

$$(\nabla^2 + k_0^2)\psi = 0 \quad \text{in} \quad D, \tag{7.10}$$

$$\psi|_S = \frac{1}{k_0^2}, \quad \psi_N|_S = 0. \tag{7.11}$$

Equations (7.10) and (7.11) imply that D is a ball, as follows from theorem 3.3 in [1]. This theorem says:

Proposition 1. *If ψ satisfies equation (7.10) and $\psi|_S = c_1$, $\psi_N|_S = c_2$, where $|c_1| + |c_2| > 0$, c_1, c_2 are constants, then D is a ball.*
    Theorem 7.0.1 is proved. □
    Proof of theorem 7.0.2. Consider the function

$$\varphi(x) = \int_S g(x, s)ds. \tag{7.12}$$

Assume that

$$\int_S e^{ik_0\alpha \cdot s}ds = 0, \quad \forall \alpha \in S^2. \tag{7.13}$$

This means that v has a spherical set of zeros. As in theorem 7.0.1 one proves that φ solves equation (7.10) and satisfies boundary conditions (7.7). The jump relation for the normal derivative yields

$$\varphi_N^+ - \varphi_N^- = 1. \tag{7.14}$$

Since $\varphi_N^- = 0$ one has $\varphi_N^+ = 1$. Thus, φ satisfies the boundary conditions

$$\varphi|_S = 0, \quad \varphi_N|_S = 1 \tag{7.15}$$

and the equation

$$(\nabla^2 + k^2)\varphi = 0 \quad \text{in} \quad D. \tag{7.16}$$

By proposition 1, D is a ball and S is a sphere.

Theorem 7.0.2 is proved. $\quad\square$

Proof of Proposition 1 is given below for convenience of the reader. It is taken from [1].

Proof of Proposition 1. For simplicity the proof is given for $D \subset \mathbb{R}^2$.

Let $r = r(s)$ be the parametric equation of S and s be the arclength of S. We assume that the origin is inside D. Then

$$\frac{dr}{ds} = t(s),$$

where $t(s)$ is the unit vector tangent to S at the point s, i.e. the point $r(s)$. Furthermore,

$$\frac{d^2r}{ds^2} = \frac{dt}{ds} = k(s)\nu(s), \tag{7.17}$$

where $k(s) \geqslant 0$ is the curvature of S at the point s and $\nu(s)$ is the unit normal to S at the point s. If S is convex then ν is directed into D. The normal ν changes direction (sign) when s passes through the points at which the convexity of S changes sign. If S is strictly convex, then $k(s) > 0 \; \forall s \in S$.

If $\psi_N = c_2 \neq 0$ and N is the unit outer normal to S, then the sign of convexity of S is unchanged. If $c_2 > 0$ then

$$\nu(s) = -N(s), \quad k(s) > 0. \tag{7.18}$$

Differentiate the boundary condition $\psi|_S = c_1$ with respect to s and get:

$$\psi_x \frac{dx}{ds} + \psi_y \frac{dy}{ds} = 0 \quad \text{or} \quad \psi_x t_1 + \psi_y t_2 = 0. \tag{7.19}$$

Differentiate again and obtain:

$$\psi_{xx} t_1^2 + 2t_1 t_2 \psi_{xy} + \psi_{yy} t_2^2 + \psi_x x_{ss} + \psi_y y_{ss} = 0. \tag{7.20}$$

Therefore,

$$\psi_{xx}t_1^2 + 2t_1t_2\psi_{xy} + \psi_{yy}t_2^2 = k(s)c_2, \quad c_2 > 0, \tag{7.21}$$

where $\nu = -N$ and

$$\psi_x x_{ss} + \psi_y y_{ss} = \nabla\psi \cdot (k(s)\nu) = -k(s)\psi_N(s) = -k(s)c_2. \tag{7.22}$$

Differential equation (7.10) can be written as

$$\psi_{xx} + \psi_{yy} = -k_0^2 c_1. \tag{7.23}$$

Denote

$$\psi_{xx}|s = p(s), \quad \psi_{xy}|s = q(s). \tag{7.24}$$

Then

$$\psi_{yy} = -p(s) - k_0^2 c_1. \tag{7.25}$$

Let A be a 2×2 symmetric matrix with the elements

$$A_{11} = p(s), \quad A_{12} = A_{21} = q(s), \quad A_{22} = -p(s) - k_0^2 c_1. \tag{7.26}$$

Equation (7.21) is rewritten as

$$(At, t) = k(s)c_2, \tag{7.27}$$

where

$$t = t_1 e_1 + t_2 e_2, \quad (t, x) = t_1 x_1 + t_2 x_2, \quad \|t\|^2 = t_1^2 + t_2^2. \tag{7.28}$$

Let us calculate eigenvalues of A. They are roots of the equation

$$\det(A - \lambda I) = 0, \tag{7.29}$$

where I is the identity operator.

Equation (7.29) can be written as

$$\lambda^2 + k^2 c_1 \lambda - p^2(s) - q^2(s) - k_0^2 c_1 p(s) = 0. \tag{7.30}$$

Therefore, the eigenvalues are

$$\begin{aligned}
\lambda_{1,2} &= -\frac{k_0^2 c_1}{2} \pm \left(\left(\frac{k_0^2 c_1}{2} \right)^2 + p^2(s) + q^2(s) + k_0^2 c_1 p(s) \right)^2 \\
&= -\frac{k_0^2 c_1}{2} \pm \left(\left(\frac{k_0^2 c_1}{2} + p(s) \right)^2 + q^2(s) \right)^{1/2}.
\end{aligned} \tag{7.31}$$

Let us calculate the corresponding eigenvectors W_1 and W_2:

$$(A - \lambda_1 I)W_1 = 0. \tag{7.32}$$

Let

$$W_1 = W_{11}e_1 + W_{12}e_2. \tag{7.33}$$

Then

$$(A_{11} - \lambda_1)W_{11} + A_{12}W_{12} = 0, \tag{7.34}$$

$$A_{21}W_{11} + (A_{22} - \lambda_1)W_{12} = 0. \tag{7.35}$$

The eigenvector W_1 we normalize by the condition $W_{11} = 1$. Then equations (7.34) and (7.26) yield

$$W_{12} = \frac{\lambda_1 - p(s)}{q(s)} := \gamma \quad \text{if} \quad q \neq 0. \tag{7.36}$$

$$\text{If } q(s) = 0 \text{ then } W_{12} = 0 \text{ and } W_1 = e_1. \tag{7.37}$$

$$\text{If } q \neq 0 \text{ then } W_1 = e_1 + \gamma e_2, \quad \|W_1\|^2 = 1 + \gamma^2. \tag{7.38}$$

Similarly, one finds

$$W_2 = -\gamma e_1 + e_2 \quad \text{if} \quad q \neq 0, \tag{7.39}$$

$$W_2 = e_2 \quad \text{if} \quad q = 0. \tag{7.40}$$

Clearly,

$$(W_j, W_p) = \delta_{jp}(1 + \gamma^2) \quad \text{if} \quad q \neq 0, \tag{7.41}$$

$$(W_j, W_p) = \delta_{jp} \quad \text{if} \quad q = 0, \tag{7.42}$$

where δ_{jp} is the Kronecker symbol. Let us express t in terms of W_1 and W_2:

$$t = h_1 W_1 + h_2 W_2, \tag{7.43}$$

where h_1 and h_2 are constant coefficients. This representation is possible since W_1 and W_2 form a basis of $\mathbb{R}^2$. The relation (7.43) is equivalent to the linear algebraic system

$$W_{11}h_1 + W_{21}h_2 = t_1, \tag{7.44}$$

$$W_{12}h_1 + W_{22}h_2 = t_2. \tag{7.45}$$

The determinant of this system is

$$\Delta = W_{11}W_{22} - W_{12}W_{21} = 1 + \gamma^2. \tag{7.46}$$

Solving system (7.44) and (7.45) one gets

$$h_1 = \frac{t_1 + \gamma t_2}{\Delta}, \quad h_2 = \frac{-\gamma t_1 + t_2}{\Delta}. \tag{7.47}$$

Note that

$$AW_j = \lambda_j W_j, \quad j = 1, 2. \tag{7.48}$$

Thus, equations (7.27), (7.43) and (7.48) yield

$$(A(h_1 W_1 + h_2 W_2), h_1 W_1 + h_2 W_2) = \lambda_1 h_1^2 \|W_1\|^2 + \lambda_2 h_2^2 \|W_2\|^2 = k(s)c_2. \tag{7.49}$$

Thus

$$\left(\lambda_1 h_1^2 + \lambda_2 h_2^2\right)\Delta = k(s)c_2. \tag{7.50}$$

From equations (7.50) and (7.47) one derives

$$\lambda_1 (t_1 + \gamma t_2)^2 + \lambda_2 (-\gamma t_1 + t_2)^2 = k(s)c_2\Delta. \tag{7.51}$$

Choose the coordinate system in which $t_1 = 1$, $t_2 = 0$. Then equation (7.51) yields

$$\lambda_1 + \gamma^2 \lambda_2 = k(s)c_2\Delta. \tag{7.52}$$

Now choose the coordinate system in which $t_1 = 0$, $t_2 = 1$. Then equation (7.51) yields

$$\lambda_1 \gamma^2 + \lambda_2 = k(s)c_2\Delta. \tag{7.53}$$

From equations (7.52) and (7.53) one gets

$$(\lambda_1 - \lambda_2)(1 - \gamma^2) = 0. \tag{7.54}$$

If $\lambda_1 = \lambda_2 := \lambda$ then it follows from equation (7.31) that

$$q(s) = 0, \quad p(s) = -\frac{k_0^2 c_1}{2}, \quad \lambda = -\frac{k_0^2 c_1}{2}. \tag{7.55}$$

Therefore,

$$\gamma = \frac{\lambda - p(s)}{q(s)} = 0, \quad \Delta = 1, \tag{7.56}$$

and equation (7.50) yields

$$\lambda(h_1^2 + h_2^2)\Delta = \lambda = k(s)c_2, \tag{7.57}$$

so

$$k(s) = -\frac{k_0^2 c_1}{2c_2}. \tag{7.58}$$

Thus, $k(s) = $ const, so S is a circle, that is, a sphere in $\mathbb{R}^2$.

If $\gamma^2 = 1$, then $\gamma = 1$ or $\gamma = -1$. Both cases are considered similarly. Assume that $\gamma = 1$. Then equation (7.51) yields

$$\lambda_1 (t_1 + t_2)^2 + \lambda_2 (-t_1 + t_2)^2 = k(s)c_2\Delta, \quad \Delta = 1 + \gamma^2 = 2. \tag{7.59}$$

Since $t_1^2 + t_2^2 = 1$, one derives from equation (7.59) that

$$\lambda_1 + \lambda_2 + (\lambda_1 - \lambda_2)2t_1t_2 = k(s)c_2\Delta. \tag{7.60}$$

Let $t_1 = 0$ and use equations (7.60) and (7.30) to get

$$k(s) = \frac{\lambda_1 + \lambda_2}{c_2\Delta} = -\frac{k^2c_1}{2c_2}. \tag{7.61}$$

Consequently, $k(s) = $ const, so S is a circle.

Proposition 1 is proved. $\qquad\qquad\qquad\qquad\qquad\qquad\qquad\qquad\qquad\square$

7.1 Summary of the results

In this chapter for the first time a symmetry problem for harmonic analysis is formulated and proved. The results are formulated in theorems 7.0.1 and 7.0.2. The proofs are based on the fundamental theorem from [1, p 15].

References

[1] Ramm A G 2019 *Symmetry Problems. The Navier-Stokes Problem* (San Rafael, CA: Morgan & Claypool)
[2] Ramm A G 2017 *Scattering by Obstacles and Potentials* (Singapore: World Scientific)

Chapter 8

Inverse scattering problem

In this chapter we discuss the following inverse scattering problem. Let

$$[\nabla^2 + k^2 n^2(x)]u = 0, \quad n^2(x) = 1 - \frac{q(x)}{k^2}, \tag{8.1}$$

in other words,

$$(\nabla^2 + k^2 - q(x))u = 0, \tag{8.2}$$

$$u = e^{ika\cdot x} + v, \quad \alpha \in S^2, \quad v_r - ikv = o\left(\frac{1}{r}\right), \quad r = |x| \to \infty. \tag{8.3}$$

One has

$$v = A(\beta, \alpha)\frac{e^{ikr}}{r} + o\left(\frac{1}{r}\right). \tag{8.4}$$

The coefficient $A(\beta, \alpha)$ is called the scattering amplitude, $k > 0$ is a constant. The scattering problem (8.2) and (8.3) is discussed in [1].

The inverse scattering problem (ISP) consists of finding $q(x)$ from the scattering data $A(\beta, \alpha)$, $\alpha, \beta \in S^2$. This problem is studied in [1, 2].

The first problem is the uniqueness of the solution to the ISP. Let compactly supported potential $q_j(x) \in L^2(D)$. Let $u_j(x, \alpha)$ be the scattering solution corresponding to the potential $q_j(x)$, generating the scattering amplitude $A_j(\beta, \alpha)$, $k = $ const, $j = 1, 2$.

Theorem 8.0.1. *If* $A_1(\beta, \alpha) = A_2(\beta, \alpha)$, $\forall \beta, \alpha \in S^2$, *then* $q_1 = q_2$.

doi:10.1088/978-0-7503-3391-7ch8

This result was originally proved by the author in 1987, see [2–7]. Its proof is based on two lemmas.

Lemma 8.0.1. (*A global perturbation formula*)

$$-4\pi\Big(A_1(\beta, \alpha) - A_2(\beta, \alpha)\Big) = \int_D \Big[q_1(x) - q_2(x)\Big]u_1(x, \alpha)u_2(x, -\beta)dx.$$

Lemma 8.0.2. *The set* $\{u_1(x, \alpha)u_2(x, \beta)\}_{\forall \alpha, \beta \in S^2}$ *is complete in* $L^2(D)$.

Proof of Lemma 8.0.1. Consider the equations for the scattering solutions:

$$u_j = u_{0j} - \int_D gq_ju_jdy, \quad g = \frac{e^{ik|x-y|}}{4\pi|x-y|}, \tag{8.5}$$

$$u_{01} = e^{ik\alpha\cdot x}, \quad u_{02} = e^{ik\beta\cdot x} \tag{8.6}$$

Let $\int := \int_D$. One has

$$\begin{aligned}
-4\pi A_1(\beta, \alpha) &= \int u(x, -\beta)q_1u_1dx = \int (u_2(x, \beta)gq_2u_2(x))q_1u_1(x)dx \\
&= \int q_1u_1(x, \alpha)u_2(x, -\beta)dx + \int gq_2u_2(x, -\beta)q_1u_1(x, \alpha)dx,
\end{aligned} \tag{8.7}$$

$$\int gq_2u_2(x, -\beta)q_1u_1(x, \alpha)dx = \int q_2u_2(x, -\beta)(u_0(x, \alpha) - u_1(x, \alpha))dx. \tag{8.8}$$

Thus

$$-4\pi A_1(\beta, \alpha) = \int (q_1 - q_2)u_1(x, \alpha)u_2(x, -\beta)dx - 4\pi A_2(-\alpha, -\beta). \tag{8.9}$$

It is known that $A_j(-\alpha, -\beta) = A_j(\beta, \alpha)$, see, e.g. [1, p 59]. Therefore, (8.10) can be written as

$$-4\pi[A_1(\beta, \alpha) - A_2(\beta, \alpha)] = \int_D (q_1 - q_2)u_1(x, \alpha)u_2(x, -\beta)dx. \tag{8.10}$$

Lemma 8.0.1 is proved. □

Proof of lemma 8.0.2. Lemma 8.0.2 was proved in [2, p 261]. The idea of the proof is based on the Property C for partial differential equations which was introduced by the author in [1–4, 6] and applied to a study of several inverse problems for elliptic, hyperbolic and parabolic equations. This property says that the set of products of solutions to some homogeneous linear partial differential equations is dense in $L^2(D)$, where $D \subset \mathbb{R}^3$ is a bounded domain, see [1, 3, 4].

The scattering solutions $u(x, \alpha)$ corresponding to a compactly supported potential is analytic with respect to $\alpha \in M$, where $M \subset \mathbb{C}^3$ is the variety defined by the equation $\sum_{j=1}^{3}\alpha_j^2 = 1$, $\alpha_j = \mathbb{C}$, see [1]. One can check that

$$u(x, \alpha) = e^{ik\alpha \cdot x}\left(1 + O\left(\frac{1}{|\alpha|}\right)\right), \quad |\alpha| \to \infty, \quad \alpha \in M. \tag{8.11}$$

Therefore,

$$u_1(x, \alpha)u_2(x, \beta) = e^{ik(\alpha+\beta)\cdot x}\left(1 + O\left(\frac{1}{|\alpha|} + \frac{1}{|\beta|}\right)\right), \tag{8.12}$$
$$|\alpha| \to \infty, \quad |\beta| \to \infty, \quad \alpha, \beta \in M$$

One can check that for any $\xi \in \mathbb{R}^3$ one can find $\alpha, \beta \in M$, $|\alpha| \to \infty$, $|\beta| \to \infty$, such that

$$\alpha + \beta = \xi, \quad \alpha, \beta \in M. \tag{8.13}$$

Thus,

$$u_1(x, \alpha)u_2(x, \beta) = e^{ik\xi \cdot x}(1 + o(1)), \tag{8.14}$$

since the set $\{e^{ik\xi \cdot x}, \forall \xi \in \mathbb{R}^3\}$ is complete in $L^2(D)$. Lemma 8.0.2 is proved. □

Proof of theorem 8.0.1. If $A_1(\beta, \alpha) = A_2(\beta, \alpha)$ then (8.5) implies

$$\int_D (q_1 - q_2)u_1(x, \alpha)u_2(x, \beta)dx = 0 \tag{8.15}$$

By lemma 8.0.2 it follows from (8.16) that $q_1 = q_2$.
Theorem 8.0.1 is proved. □

Exercise 8.1. *Check that for any $\xi \in \mathbb{R}^3$ there exist $\alpha, \beta \in M$, $|\alpha| \to \infty$, $|\beta| \to \infty$, such that $\alpha + \beta = \xi$.*

Solution. If $\alpha, \beta \in M$ then

$$\sum_{j=1}^{3}\alpha_j^2 = 1, \quad \alpha_j \in \mathbb{C}, \tag{8.16}$$

$$\sum_{j=1}^{3}\beta_j^2 = 1, \quad \beta_j \in \mathbb{C}. \tag{8.17}$$

Let $\alpha_j = a_j + ib_j$, $\beta_j = p_j + iq_j$, $a_j, b_j, p_j, q_j \in \mathbb{R}^1$. Then equations (8.17) and (8.18) hold if

$$\sum_{j=1}^{3} a_j^2 - b_j^2 = 1, \quad \sum_{j=1}^{3} a_j b_j = 0, \tag{8.18}$$

$$\sum_{j=1}^{3} p_j^2 - q_j^2 = 1, \quad \sum_{j=1}^{3} p_j q_j = 0. \tag{8.19}$$

Let e_m, $m = 1, 2, 3$, be an orthonormal basis of $\mathbb{R}^3$. Then vectors $a = \sum_{j=1}^{3} a_j e_j$ and $b = \sum_{j=1}^{3} b_j e_j$ are orthogonal, $|a|^2 - |b|^2 \geq 1$ and so are vectors p and q, and $p = \sum_{j=1}^{3} p_j e_j$, $q = \sum_{j=1}^{3} a_j e_j$, $p \cdot q = 0$, $|p|^2 - |q|^2 = 1$.

Also

$$a + ib + p + iq = \xi \tag{8.20}$$

implies

$$a + p = \xi, \quad \text{so} \quad a_j + p_j = \xi_j, j = 1, 2, 3, \tag{8.21}$$

$$b + q = 0, \quad \text{so} \quad b_j = -q_j. \tag{8.22}$$

Choose $\{e_m\}_{m=1}^{3}$ so that

$$\xi_1 = a_1 e_1, \quad \xi_2 = \xi_3 = 0. \tag{8.23}$$

Then conditions (8.19)–(8.21) hold if, for example,

$$a = a_1 e_1, \quad b = b_2 e_2 + b_3 e_3, \tag{8.24}$$

and

$$a_1^2 - b_2^2 - b_3^2 = 1, \tag{8.25}$$

$$p_1^2 - q_2^2 - q_3^2 = 1, \tag{8.26}$$

$$a_1 + p_1 = \xi_1. \tag{8.27}$$

Equation (8.28) holds if $p_1 = \xi_1 - a_1$. Equation (8.27) holds if

$$(\xi_1 - a_1)^2 - q_2^2 - q_3^2 = 1. \tag{8.28}$$

Conditions

$$a_1^2 + b_2^2 + b_3^2 \to \infty, \quad p_1^2 + q_2^2 + q_3^2 \to \infty, \tag{8.29}$$

hold if $b_2^2 + b_3^2 \to \infty$, $q_2^2 + q_3^2 \to \infty$, $a_1^2 \to \infty$, $p_1^2 \to \infty$. Conditions (8.26) and (8.29) hold, which is clearly possible to have by infinitely many choices of b_2, b_3, q_2, q_3 and a_1. □

8.1 Summary of the results

In this chapter the inverse scattering problem with fixed-energy data is discussed. The basic result is theorem 8.0.1 which was proved by the author in 1987. The notion of Property C is introduced and applied, following the author's original proof.

References

[1] Ramm A G 2017 *Scattering by Obstacles and Potentials* (Singapore: World Scientific)
[2] Ramm A G 2005 *Inverse Problems* (New York: Springer)
[3] Ramm A G 1987 Completeness of the products of solutions to PDE and uniqueness theorems in inverse scattering *Inverse Problems* **3** L77–82
[4] Ramm A G 1988 Multidimensional inverse problems and completeness of the products of solutions to PDE *J. Math. Anal. Appl.* **134** 211–53
Ramm A G 1989 Multidimensional inverse problems and completeness of the products of solutions to PDE *J. Math. Anal. Appl.* **139** 302
[5] Ramm A G 1988 Recovery of the potential from fixed energy scattering data *Inverse Problems* **4** 877–86
Ramm A G 1989 Recovery of the potential from fixed energy scattering data *Inverse Problems* **5** 255
[6] Ramm A G 1990 Completeness of the products of solutions of PDE and inverse problems *Inverse Problems* **6** 643–64
[7] Ramm A G 1992 *Multidimensional Inverse Scattering Problems* (New York: Longman)

IOP Publishing

Creating Materials with a Desired Refraction Coefficient (Second Edition)

Alexander G Ramm

Bibliography

[1] Ramm A G 2007 Materials with the desired refraction coefficients can be made by embedding small particles *Phys. Lett.* A **370** 522–7

[2] Ramm A G 2007 Distribution of particles which produces a desired radiation pattern *Commun. Nonlinear Sci. Numer. Simul.* **12** 1115–9

[3] Ramm A G 2007 Distribution of particles which produces a 'smart' material *J. Stat. Phys.* **127** 915–34

[4] Ramm A G 2007 Distribution of particles which produces a desired radiation pattern *Physica* B **394** 253–5

[5] Ramm A G 2008 A recipe for making materials with negative refraction in acoustics *Phys. Lett.* A **372** 2319–21

[6] Ramm A G 2008 Wave scattering by many small particles embedded in a medium *Phys. Lett.* A **372** 3064–70

[7] Ramm A G 2009 Preparing materials with a desired refraction coefficient *Nonlinear Anal.: Theory Methods Appl.* **70** e186–90

[8] Ramm A G 2010 A method for creating materials with a desired refraction coefficient *Int. J. Mod. Phys.* B **24** 5261–8

[9] Ramm A G 2011 Wave scattering by many small bodies and creating materials with a desired refraction coefficient *Afr. Mat.* **22** 33–55

[10] Ramm A G 2011 Electromagnetic wave scattering by a small impedance particle of arbitrary shape *Opt. Commun.* **284** 3872–7

[11] Ramm A G 2013 Many-body wave scattering problems in the case of small scatterers *J. Appl. Math. Comput.* **41** 473–500

[12] Ramm A G and Tran N 2015 A fast algorithm for solving scalar wave scattering problem by billions of particles *J. Algorithms Optim.* **3** 1–13

[13] Ramm A G 2015 Scattering of EM waves by many small perfectly conducting or impedance bodies *J. Math. Phys.* **56** 091901

[14] Ramm A G 2015 EM wave scattering by many small impedance particles and applications to materials science *Open Opt. J.* **9** 14–7

[15] Ramm A G 2018 Many-body wave scattering problems for small scatterers and creating materials with a desired refraction coefficient *Mathematical Analysis and Applications: Selected Topics* ed M Ruzhansky, H Dutta and R Agarwal (Hoboken, NJ: Wiley) ch 3, pp 57–76

[16] Ramm A G 2010 On a hyper-singular equation *Open J. Math. Anal.* **4** 8–10

[17] Ramm A G 2018 Existence of the solution to convolution equations with distributional kernels *Global J. Math. Anal.* **6** 1–2

Lightning Source UK Ltd.
Milton Keynes UK
UKHW052033290720
367351UK00003B/49

9 780750 333894